T0297258

LONDON MATHEMATICAL SOCIETY STUDENT TEXTS

Managing Editor: Professor D. Benson,
Department of Mathematics, University of Aberdeen, UK

London Mathematical Society Student Texts 77

Lectures on Profinite Topics in Group Theory

BENJAMIN KLOPSCH
Royal Holloway, University of London

NIKOLAY NIKOLOV
Imperial College London

CHRISTOPHER VOLL
University of Southampton

edited by

DAN SEGAL
All Souls College, Oxford

CAMBRIDGE
UNIVERSITY PRESS

CAMBRIDGE
UNIVERSITY PRESS

Shaftesbury Road, Cambridge CB2 8EA, United Kingdom

One Liberty Plaza, 20th Floor, New York, NY 10006, USA

477 Williamstown Road, Port Melbourne, VIC 3207, Australia

314–321, 3rd Floor, Plot 3, Splendor Forum, Jasola District Centre, New Delhi – 110025, India

103 Penang Road, #05–06/07, Visioncrest Commercial, Singapore 238467

Cambridge University Press is part of Cambridge University Press & Assessment, a department of the University of Cambridge.

We share the University's mission to contribute to society through the pursuit of education, learning and research at the highest international levels of excellence.

www.cambridge.org
Information on this title: www.cambridge.org/9780521183017

First published 2011

A catalogue record for this publication is available from the British Library

Library of Congress Cataloging-in-Publication data
Klopsch, Benjamin.
Lectures on profinite topics in group theory / Benjamin Klopsch, Nikolay Nikolov, and Christopher Voll ; edited by Dan Segal.
p. cm.
ISBN 978-1-107-00529-7 (Hardback) – ISBN 978-0-521-18301-7 (pbk.)
1. Profinite groups. 2. Group theory. I. Nikolov, Nikolay. II. Voll, Christopher. III. Segal, Daniel, 1947– IV. Title.
QA177.K56 2011
512´.2–dc22

2010046477

ISBN   978-1-107-00529-7   Hardback
ISBN   978-0-521-18301-7   Paperback

Contents

Preface

In September 2007, the London Mathematical Society and the EPSRC sponsored a 'short course for graduates' in Oxford, under the heading 'Asymptotic methods in infinite group theory'. This was organised by Dan Segal and consisted of three series of lectures. The present book is basically a record of these lectures, somewhat polished and expanded. It is intended to serve as an introduction, for beginning research students and for interested non-specialists, to some areas of current activity in algebra: the questions mostly originate in group theory but the methodology encompasses a wide range of mathematics, involving topology, algebraic geometry, number theory and combinatorics.

Editor's introduction

From a purely algebraic point of view, there is not a lot one can say about infinite groups in general. Traditionally, these have been studied to good effect in combination with topology or geometry. These lectures represent an introduction to some recent developments that arise out of looking at infinite groups from a point of view inspired – in a general sense – by number theory; specifically the interaction between 'local' and 'global', where by 'local' properties of a group G, in this context, one means the properties of its finite quotients, or equivalently properties of its profinite completion $\widehat{G}$. The second chapter directly addresses the interplay between certain finitely generated groups and their finite images. The other two chapters are more specifically 'local' in emphasis: Chapter I concerns the algebraic structure of certain pro-p groups, while Chapter III introduces a way of studying the rich arithmetical data encoded in certain infinite groups and related structures.

A motivating example for all of the above is the question of 'subgroup growth'. Say G has $s_n(G)$ subgroups of index at most n for each n; the function $n \mapsto s_n(G)$ is the *subgroup growth function* of G, and is finite-valued if we assume that G is finitely generated. Now we can ask (inspired perhaps by Gromov's celebrated *polynomial growth theorem*): what does it mean for the global structure of a finitely generated group if its subgroup growth function is (bounded by a) polynomial? To approach a question of this kind, we need to show that if G is in some sense very big, or very complicated, then G must have a lot of finite quotients that can be more or less well understood. If G is a finitely generated *linear group*, there is a natural family of such quotients provided by the *congruence subgroups*. The theory of 'strong approximation' gives remarkably good information about these; this is the topic of Chapter II.

The point of 'local–global' results in number theory is that the 'local' situation is usually easier to understand. In group theory, we can similarly make things easier by restricting attention to p-groups: there is only one finite simple p-group! To an infinite group G we can associate its pro-p completion $\widehat{G}_p$, which is the inverse limit of the finite p-group quotients of G. If these finite p-quotients are suitably 'small' (for example, if G has polynomial subgroup growth), then – wonderfully! – $\widehat{G}_p$ turns out to have the structure of a (p-adic) Lie group. This has manifold consequences; in particular, $\widehat{G}_p$ is a linear group. Thus the natural map from G into $\widehat{G}_p$ provides a linear representation of G, and the whole technology of linear groups (including the methods of Chapter II) can be applied. Of course, p-adic Lie groups arise in many other situations. Chapter I presents an elementary introduction to the topic, from a group-theoretic perspective.

Given a group G, we can also study the arithmetic of the sequence $(s_n(G))$, or of other sequences associated to G in a similar spirit (such as the *representation growth* function). If G is not 'too big' – finitely generated and nilpotent, say, or arithmetic, or p-adic analytic – these sequences have amazing properties. This is the topic of Chapter III, which introduces the *zeta functions* attached to certain groups and rings. This is a subject still in its infancy: while many striking results have been obtained, many tantalising questions remain.

The three chapters can be read independently of one another, though there are occasional cross-references; for a quick introduction to p-adic numbers and profinite groups see Sections 2–5 of Chapter I. Each chapter has its own introduction; the following remarks are more by way of general motivation.

Analytic pro-p groups

If we want to study the finite images of a group like $\mathrm{SL}_n(\mathbb{Z})$ from a 'local' point of view, we may focus on those of the form $\mathrm{SL}_n(\mathbb{Z}/p^m\mathbb{Z})$ for a fixed prime p. The inverse limit of this system is the group $\mathrm{SL}_n(\mathbb{Z}_p)$ (where $\mathbb{Z}_p$ is the ring of p-adic integers). This is the prototype of a (compact) *p-adic analytic group*. As one would hope, its structure is more transparent than that of the original arithmetic group $\mathrm{SL}_n(\mathbb{Z})$. In particular, it has an open (finite-index) normal subgroup which is a pro-p group of finite rank. In general, one obtains such a pro-p group as an inverse limit of finite p-groups of uniformly bounded ranks; Chapter I presents some of the rich structural theory that exists for these groups. This material belongs in every group-theorist's toolbox.

The chapter introduces the concept of pro-p groups, as inverse limits of finite p-groups. It then develops in more detail the theory of pro-p groups of finite rank – in this context, the *rank* of a group G can be defined as the largest dimension (over $\mathbb{F}_p$) of an elementary abelian section of G. If this is finite, it turns out that G (or at least a suitable subgroup of finite index) carries the structure of a *Lie algebra* over the p-adic integers $\mathbb{Z}_p$, of the same (*finite*) dimension. Thus certain questions about the non-commutative group G can be approached with the help of linear methods.

One consequence is that a pro-p group G of finite rank has the structure of an *analytic group* over $\mathbb{Q}_p$. The more analytic aspects of the theory are not explored in depth in this chapter; a fuller account may be found in the book [APG]. Here it is pointed out that a p-adic analytic pro-p group is the same thing as a closed (in the p-adic topology) subgroup of $\mathrm{GL}_n(\mathbb{Z}_p)$, for some n, a fact that has useful applications as mentioned above.

The Lie theory is applied to good effect in studying the *finite* representations of these groups; the *Kirillov orbit method* relates these to the adjoint action of the group on its Lie algebra, and leads to remarkable results concerning the 'representation growth' functions. These in turn can be applied, in a local–global spirit, to the representation growth of arithmetic groups: a topic touched on in Chapters I and III, and the subject of much ongoing research.

Chapter I has some overlap with the book [APG], to which it may serve as an introduction; it also pursues in some depth topics not covered by that book – these include saturable pro-p groups, potent filtrations, and the Kirillov orbit method.

Strong approximation

If one wants to study linear groups, one needs to have some basic familiarity with the theory of linear algebraic groups. One purpose of Chapter II is to provide a brief overview of some of the essential features of this theory – at least enough so that the newcomer can make sense of, and appreciate the value of, the 'strong approximation' results that form the main focus.

In algebraic number theory, the Strong Approximation Theorem is a slightly beefed-up version of the Chinese Remainder Theorem, which says that if $\mathfrak{a}_1, \ldots, \mathfrak{a}_s$ are (finitely many) pairwise coprime ideals in a ring of algebraic integers $\mathfrak{o}$, then the natural map from $\mathfrak{o}$ into $\mathfrak{o}/\mathfrak{a}_1 \times \cdots \times \mathfrak{o}/\mathfrak{a}_s$ is surjective. A much deeper fact is that an analogous statement is true for certain non-commutative matrix groups (arithmetic groups). The general setup is explained in Chapter II; as a typical example, we have: if $q_1, \ldots, q_s$ are pairwise coprime integers, then the natural map $\pi : \mathrm{SL}_n(\mathbb{Z}) \to \mathrm{SL}_n(\mathbb{Z}/q_1\mathbb{Z}) \times \cdots \times \mathrm{SL}_n(\mathbb{Z}/q_s\mathbb{Z})$ is surjective.

This theory is satisfying, and in a sense not surprising ($\mathrm{SL}_n(\mathbb{Z})$ is generated by elementary subgroups that look like $\mathbb{Z}$, to which the Chinese Remainder Theorem may be applied; the proof for other arithmetic groups is much harder). A truly remarkable generalisation was discovered in the 1980s by Madhav Nori and Boris Weisfeiler. This applies to linear groups that may be far from arithmetic; for example, if Γ is *any* Zariski-dense subgroup of $\mathrm{SL}_n(\mathbb{Z})$, then the restriction of π to Γ is still surjective, as long as $q_1, \ldots, q_s$ avoid some finite set of possibly bad primes. In general, the theorem applies to any linear group Γ (over a ring $\mathbb{Z}[1/m]$ for some m) such that the Zariski-closure G of Γ is simple as an algebraic group over $\mathbb{Q}$: it ensures that Γ has an infinite family of readily identifiable finite images, namely the groups $G(\mathbb{Z}/q\mathbb{Z})$ for many integers q.

The necessary technical language (algebraic groups, Zariski topology) is all explained in this chapter, which goes on to describe some powerful and straightforward applications to finitely generated linear groups in general. These are encapsulated in the so-called 'Lubotzky alternative', which implies the following: if Γ is a finitely generated linear group over a field of characteristic 0, then *either* Γ is virtually soluble *or* Γ has a subgroup Δ of finite index such that Δ has infinitely many finite quotients of the form $G(\mathbb{F}_{p^e})$, simple groups of a fixed Lie type over finite fields, with p ranging over almost all primes and e bounded.

The Lubotzky alternative is a key tool for attacking questions of the following kind (and was indeed motivated by them): what global constraints follow for a finitely generated group G if the finite images of G are in some sense 'small', or in some sense 'grow slowly'? Such investigations are expounded in Chapters 5 and 12 of the book [SG]; results include the characterisation of finitely generated residually finite groups of finite upper rank, or with polynomial subgroup growth, as those which are virtually soluble of finite rank. The point in each case is that if G is not virtually soluble, then G must have finite images that are 'too big', or grow too fast.

This methodology has been used in a number of other ways (see Chapter II, Section 6.4). It also belongs in the toolbox of anyone seriously studying finitely generated residually finite groups.

Zeta functions

To each finitely generated group G we may associate the numerical sequence (a_n) where $a_n = a_n(G)$ is the number of subgroups of index (exactly) n in G. It is traditional in number theory to represent such a sequence by a 'generating function'. If the a_n grow at most polynomially with n (i.e. if G has *polynomial subgroup growth*), it may be a good idea to take for this the Dirichlet series

$$\zeta_G(s) = \sum_{n=1}^{\infty} a_n n^{-s}.$$

This is a priori a formal expression in which s is an indeterminate, but the polynomial growth condition implies (and is equivalent to) the fact that $\zeta_G(s)$ converges if s is a complex number lying in some non-empty right half-plane, and defines there an analytic function of s. A familiar example is where G is the infinite cyclic group, in which case ζ_G is the Riemann zeta function $\zeta(s)$. Of course, the sequence – $a_n = 1$ for all n – encoded by $\zeta(s)$ does not in itself seem very challenging. However, if instead of $\mathbb{Z}$ we consider the ring of integers $\mathfrak{o}$ in an algebraic number field k and let a_n denote the number of ideals of index (i.e. *norm*) n in $\mathfrak{o}$, the resulting Dirichlet series is then the Dedekind zeta function ζ_k; over a century of algebraic and analytic number theory has shown how the analysis of ζ_k reveals deep properties of the number field k.

The number-theoretic zeta functions have many excellent properties, such as an Euler product, analytic continuation, functional equations. It would be too much to expect all of these to obtain if we start from an essentially non-commutative object like a finitely generated (non-abelian) group. However, for certain kinds of group the associated zeta functions turn out to have some remarkable properties: for example, if G is nilpotent then ζ_G does have an Euler product. This is not so surprising; more remarkably, the 'local factor' at a prime p is a *rational function* in the parameter p^{-s} (recall that the p-local factor of the Riemann zeta function is $1/(1 - p^{-s})$). We can then seek more detailed information about these rational functions: are they all the same (as in Riemann's case)? What other properties do they have? It turns out that the Dedekind zeta functions are not quite an adequate model: more relevant are the Hasse–Weil zeta functions associated to algebraic varieties.

Various zeta functions of this general nature can be associated to various kinds of groups and rings. Chapter III introduces some of these, and presents methods used for analysing them. The non-commutative nature of the input means that, in all but the simplest cases, the explicit calculation of the functions is very hard. Many remarkable results have nonetheless been achieved. Among the most remarkable is the widespread occurrence of so-called 'local functional equations'. This was quite unexpected, remaining for a long time no more than a collection of experimental observations. It reveals deep hidden arithmetical symmetries in apparently innocuous algebraic structures (partly related to – though not a simple consequence of – the Weil conjectures).

Chapter III is perhaps the most technically demanding part of the book: it serves as an introduction to a rich field of research that is only beginning to reveal its mysteries. Less technical – and less up-to-date – discussions of these topics can be found in Chapter 9 of [NH] and Chapters 15 and 16 of [SG].

References for Chapter

[APG] J. D. Dixon, M. P. F. du Sautoy, A. Mann and D. Segal, *Analytic pro-p groups, 2nd edn*, Cambridge University Press, 1999.

[NH] M. P. F. du Sautoy, D. Segal and A. Shalev (eds.), *New horizons in pro-p groups*, Birkhäuser, Boston MA, 2000.

[SG] A. Lubotzky and D. Segal, *Subgroup growth*, Birkhäuser, Basel, 2003.

Chapter I

An introduction to compact p-adic Lie groups

by Benjamin Klopsch

1 Introduction

The theory of Lie groups is highly developed and of relevance in many parts of contemporary mathematics and theoretical physics. Loosely speaking, a Lie group is a group with the additional structure of a real differentiable manifold, given by local coordinate systems, such that the group operations are smooth functions.

Historically, the study of Lie groups, over the real and complex numbers, arose toward the end of the 19th century, from the analysis of continuous symmetries of differential equations by the mathematician Sophus Lie and others. Around the middle of the 20th century, mathematicians such as Armand Borel and Claude Chevalley found that many of the foundational results concerning Lie groups could be developed completely algebraically, giving rise to the theory of algebraic groups defined over arbitrary fields. This insight opened the way for entirely new directions of investigation. Much of the theory of p-adic Lie groups was developed in the 1960s by mathematicians such as Nicolas Bourbaki, Michel Lazard and Jean-Pierre Serre. Since then the study of p-adic Lie groups and analogues of Lie groups over adele rings has largely been motivated by questions from number theory, e.g. regarding automorphic forms and Galois representations. More recently, p-adic Lie groups have also become a key tool in infinite group theory.

Throughout, let p be a prime. The real numbers $\mathbb{R}$ form a completion of the rational numbers $\mathbb{Q}$. Similarly, the field of p-adic numbers $\mathbb{Q}_p$ is obtained by completing $\mathbb{Q}$, albeit with respect to a different, non-archimedean notion of distance. One can define analytic functions over $\mathbb{Q}_p$ and p-adic manifolds, just as over $\mathbb{R}$. A p-adic Lie group, or p-adic analytic group, is a Lie group whose local coordinate systems are p-adic valued rather than real valued. Given such a group, the usual apparatus of Lie theory is available; but one needs to keep in mind that the underlying geometry is rather different, i.e. non-archimedean.

Based on [6, Historical Note, VII], we give some indication of the early history of p-adic Lie theory. The first p-adic Lie groups were encountered by Kurt Hensel at the beginning of the 20th century. He was interested in the local isomorphisms between the additive and the multiplicative groups of $\mathbb{Q}_p$, via the exponential and logarithm maps. More general commutative p-adic Lie groups appeared in the works of André Weil and Élisabeth Lutz on elliptic curves in the 1930s. Subsequent investigations of abelian varieties by Claude Chabauty suggested that the local theory of Lie groups could be applied with little change to the p-adic setting. In 1942, this was made explicit by Robert Hooke, a student of Chevalley; see [19]. Until the beginning of the 1960s, p-adic Lie theory continued to be of interest mainly to arithmeticians and algebraic geometers.

The crucial turning point came in 1962, when Jean-Pierre Serre was prompted by a question of John Tate to consider the cohomology of a closed subgroup of the p-adic Lie group $\mathrm{GL}_2(\mathbb{Z}_p)$. His work led him to propose to Michel Lazard a general programme of comparing the cohomology of p-adic Lie groups to the cohomology of associated Lie algebras.[1] In addition to his cohomological results, Lazard's great achievement in [39] was to show that the class of p-adic Lie groups admits a fairly straightforward group-theoretic characterisation, thereby solving the p-adic analogue of Hilbert's fifth problem.

The upshot of Lazard's characterisation and its later interpretation in terms of powerful groups and groups of finite rank, as described in [10], is that one can study and utilise compact p-adic Lie groups without ever imposing any analytic machinery. Instead, one can construct internally, by group-theoretic means, the key features and invariants of such groups, e.g. their dimensions as Lie groups. This truly algebraic nature of p-adic Lie groups explains to a certain degree their continuing relevance and usefulness in infinite group theory throughout the last three decades; e.g. see [10] and the references given therein.

It is very natural to ask to what extent this success story also translates to groups which are analytic over local fields of positive characteristic or, more generally, pro-p domains of higher Krull dimension. Here our understanding is still much less complete; cf. [10, Ch. 13] and [25].

Aims and scope

The aim of the present notes is to provide an accessible introduction to compact p-adic Lie groups from a group-theoretic point of view. We also discuss the relation between p-adic analytic pro-p groups, other classes of profinite groups and abstract groups. The text is based on a series of five lectures delivered during a short course for graduate students at the University of Oxford in 2007. I have tried to preserve the basic structure and informal style of the original lectures, while adding slightly more detail and appropriate references in places. The series

---

[1] We are grateful to Prof. Serre for providing this historical information.

of exercises which I include are essentially the ones given during the course, and one of their key purposes is to explore topics which branch off from the main thread of narration. Readers of these notes who are subsequently interested in a more detailed account of the theory of p-adic analytic pro-p groups will naturally turn to the book [10], by Dixon, du Sautoy, Mann and Segal, and the other books listed as main references below.

Content and organisation

The notes are organised as follows.

Section 2 provides a short account of prerequisites from group theory, algebra and number theory. The main topics discussed are: nilpotent groups, finite p-groups, Lie rings, Lie methods in group theory, absolute values, p-adic numbers and integers. The section ends with a preview of what p-adic analytic groups are. The short Section 3 provides a summary of basic notions and facts from point-set topology. Section 4 contains the first series of exercises.

Section 5 introduces powerful finite p-groups and profinite groups (as Galois groups, inverse limits, profinite completions and topological groups). It goes on to describe pro-p groups, powerful pro-p groups and pro-p groups of finite rank. The latter are precisely the pro-p groups which admit the structure of a p-adic analytic group. The second series of exercises is collected in Section 6.

Section 7 describes uniformly powerful pro-p groups and the powerful $\mathbb{Z}_p$-Lie lattices associated to them. Both directions, the limit process which yields a Lie lattice from a Lie group and the transition from a Lie lattice to a Lie group via the Hausdorff formula are explained.

Section 8 starts with a concrete example, the group $\mathrm{GL}_d(\mathbb{Z}_p)$ and its principal congruence subgroups. It then moves on to discuss just-infinite pro-p groups, saturable pro-p groups and the Lie correspondence between subgroups of saturable pro-p groups and Lie sublattices of the associated $\mathbb{Z}_p$-Lie lattice. The third and last series of exercises is collected in Section 9.

Section 10 provides a taste of current research on complex irreducible representations of compact p-adic Lie groups. It introduces the Kirillov orbit method and illustrates its use in the study of representation zeta functions.

References

The following books, which can be regarded as our main references, cover some of the selected material in greater detail. They also address many related and more advanced topics.

J. D. Dixon, M. P. F. du Sautoy, A. Mann and D. Segal, *Analytic pro-p groups*, Cambridge University Press, 1999.

E. I. Khukhro, *p-automorphisms of finite p-groups*, Cambridge University Press, 1998.

G. Klaas, C. R. Leedham-Green and W. Plesken, *Linear pro-p-groups of finite width*, Springer Verlag, 1997.

C. R. Leedham-Green and S. McKay, *The structure of groups of prime power order*, Oxford University Press, 2002.

J. S. Wilson, *Profinite groups*, Oxford University Press, 1998.

The original source for much of the theory of p-adic analytic groups is Lazard's seminal paper 'Groupes analytiques p-adiques', Inst. Hautes Études Scientifiques, *Publ. Math.* 26, 389–603 (1965).

Throughout the text I have aimed to give reasonably complete, but not exhaustive references to the literature. A guiding principal for my choices has been to select economically a mixture of classical and modern references which are suitable for a newcomer to the subject. More complete references can be found in the books listed above. Each section of the present notes, except for the short Section 3, ends with a few selected suggestions for further reading.

Acknowledgements

In preparing the original course and these notes, I made considerable use of several of the main references listed above, in particular the first book. I also included key results from selected research articles and preprints. Originality I can claim, in a limited sense, with regard to the overall exposition. I am grateful to Dan Segal, Christopher Voll and the anonymous referees for their comments on earlier versions of this text. Given the informal style of the notes I made a fair, but perhaps not entirely systematical effort to attribute results to their respective authors; I apologise for any shortcomings of this light approach.

2   From finite p-groups to compact p-adic Lie groups

In this section, we provide a short account of various basic concepts from group theory and number theory, and we introduce some key notation. After discussing finite p-groups, Lie methods and p-adic integers, we state a hands-on version of Lazard's characterisation of compact p-adic Lie groups.

A useful, general reference for the group-theoretic notions and facts, appearing in this section, is Robinson's introductory text [51].

2.1   Nilpotent groups

Let G be a group and let $x, y \in G$. The *conjugate* of x by y is $x^y = y^{-1}xy$. Conjugation provides a natural action of G on itself; indeed, it induces a homomorphism from G into its automorphism group $\mathrm{Aut}(G)$. The kernel of this homomorphism, which constitutes a normal subgroup of G, is called the *centre* of G and denoted by $\mathrm{Z}(G)$. The *upper central series* of G is the ascending series of normal subgroups

$$1 = \mathrm{Z}_0(G) \leq \mathrm{Z}_1(G) \leq \ldots, \quad \text{where } \mathrm{Z}_{i+1}(G)/\mathrm{Z}_i(G) = \mathrm{Z}(G/\mathrm{Z}_i(G)).$$

By and large, we will be interested in filtrations of a group G which start at the top, such as the lower central series which we describe next. The *commutator* of x with y is $[x,y] = x^{-1}x^y = x^{-1}y^{-1}xy$. The subgroup generated by all commutators is called the *commutator subgroup* of G and denoted by $[G,G]$. This notation is easily adapted to a more general situation: if $H, K \leq G$, then we write $[H,K]$ to denote the subgroup of G which is generated by all commutators $[h,k]$ with $h \in H$ and $k \in K$. The group $[G,G]$ can be characterised as the smallest normal subgroup of G such that the corresponding quotient is abelian. The *lower central series* of G is the descending series of normal subgroups

$$G = \gamma_1(G) \geq \gamma_2(G) \geq \ldots, \quad \text{where } \gamma_{i+1}(G) = [\gamma_i(G), G].$$

A basic property of this sequence is that $[\gamma_i(G), \gamma_j(G)] \subseteq \gamma_{i+j}(G)$ for all $i, j \in \mathbb{N}$.

The group G is said to be *nilpotent* if its lower central series terminates in the trivial group 1 after finitely many steps; in this case, the nilpotency class of G is the smallest non-negative integer c such that $\gamma_{c+1}(G) = 1$. It can be shown that for any group G and for any natural number c the lower central series of G terminates in 1 after c steps if and only if the upper central series of G terminates in G after c steps; see [51, §5.1.9].

Nilpotent groups can be thought of as close relatives of abelian groups. Nevertheless, the study of finite nilpotent groups can become exceedingly difficult from a purely group-theoretic point of view. In fact, a finite group is nilpotent if and only if for each prime p it has a unique Sylow p-subgroup. Equivalently, a finite group is nilpotent if and only if it decomposes as a direct product of finite p-groups; see [51, §5.2.4]. Whereas finite abelian groups are completely classified, the theory of finite p-groups remains an active area of research with many open problems.

Of particular interest in finite group theory is the information that can be gained about a group G from its Sylow p-subgroups – which, as indicated, are nilpotent – and their normalisers. This direction, called *local group theory*, played a critical role in the classification of finite simple groups. For instance, relating the representation theory of a finite group G to the representation theory of the normalisers of p-subgroups of G is currently an attractive field of research; a lot of recent work is focused around the McKay conjecture and generalisations thereof, e.g. see [46].

2.2 Finite p-groups

A *p-group* is a torsion group in which every element has p-power order. Accordingly, finite p-groups are precisely the groups of p-power order. We implicitly stated above that every finite p-group is nilpotent. This fact can easily be proved inductively from the following fundamental observation. Every non-trivial normal subgroup N of a finite p-group G intersects $Z(G)$ non-trivially. In particular, the centre of a non-trivial finite p-group is non-trivial. This observation can be proved by analysing the possible orbit sizes in the action of G on N by conjugation; see Exercise 4.1. An interesting consequence is that every proper subgroup of a finite p-group G is properly contained in its normaliser.

It is easy to see that the maximal subgroups of a finite p-group G are precisely the subgroups of index p and hence normal in G. The intersection of all maximal subgroups of G is the *Frattini subgroup*, commonly denoted by $\Phi(G)$. The factor group $G/\Phi(G)$ constitutes the largest elementary abelian quotient of the finite p-group G. In other words, the Frattini subgroup can be described as $\Phi(G) = G^p[G, G]$, where G^p denotes the subgroup generated by all pth powers in G.

The Frattini subgroup of a finite p-group G plays a useful role in the context of generating sets. Let $X \subseteq G$. Then X generates G if and only if there is no maximal subgroup of G containing X. This shows that X generates G if and only if its image modulo $\Phi(G)$ constitutes a generating set of $G/\Phi(G)$. Being an elementary p-group, $G/\Phi(G)$ can be regarded as a finite-dimensional vector space V over the finite prime field $\mathbb{F}_p$. The set X is a minimal generating set of G if and only if its image in V forms a basis for V. Thus all minimal generating sets of G have the same size, namely $\dim_{\mathbb{F}_p} V$.

Generating sets for finite groups which are not p-groups and not nilpotent are more difficult to understand and can be of considerable combinatorial interest; for instance, see [41] for sharp bounds on the diameters of finite simple groups. 'Efficient' presentations – for instance, for finite simple groups – are also of great importance from a computational point of view.

2.3 Lie rings

Lie methods constitute an important tool in the study of groups. In particular, this applies to p-groups and, more generally, pro-p groups. The basic idea is to capture a large part of the group structure in a Lie ring.

We recall that a *Lie ring* is a $\mathbb{Z}$-module L together with a bi-additive operation $[\cdot, \cdot] : L \times L \to L$ which is skew-symmetric and satisfies the Jacobi identity

$$[x, x] = 0 \quad \text{and} \quad [[x, y], z] + [[y, z], x] + [[z, x], y] = 0 \quad \text{for all } x, y, z \in L.$$

Let R be a commutative ring, the most common case being that R is a field. If L has the additional structure of an R-module and if $[\cdot, \cdot]$ is bilinear with respect to scalar multiplication by elements of R, then L is called a *Lie algebra* over R. If R is a principal ideal domain and L is a free R-module of finite rank, one also uses the term *Lie lattice*. Standard examples of Lie algebras include matrix algebras. Let $d \in \mathbb{N}$. Then the set $\mathfrak{gl}_d(R)$ of $d \times d$ matrices over R, regarded as an R-module and endowed with the commutator bracket

$$[A, B] := AB - BA \quad \text{for all } A, B \in \mathfrak{gl}_d(R)$$

forms a Lie algebra over R. In fact, a theorem of Ado states that every finite-dimensional Lie algebra over a field K of characteristic 0 is isomorphic to a Lie subalgebra of $\mathfrak{gl}_d(K)$ for a suitable degree d; see [6, I.§7].

At first sight, Lie algebras perhaps appear to be more complicated objects than groups. However, one should think of a Lie algebra essentially as a vector space. The extra structure, given by the Lie bracket, can be regarded as a simplified version of the group commutator. For instance, the group-theoretic analogue of the Jacobi identity is the 'baffling' Hall–Witt identity

$$[[x, y^{-1}], z]^y \, [[y, z^{-1}], x]^z \, [[z, x^{-1}], y]^x = 1$$

which holds in any group; see [51, §5.1.5]. Many of the concepts we have introduced for groups, such as nilpotency, can be defined mutatis mutandis in the context of Lie rings. For instance, the centre of a Lie ring L is the Lie ideal $Z(L) = \{x \in L \mid \forall y \in L : [x, y] = 0\}$. We trust that the reader will make the appropriate translations of this kind where necessary; e.g. see [29, Ch. 5].

2.4 Applying Lie methods to groups

Next we describe a comparatively simple recipe for associating a Lie ring to a group G with respect to its lower central series. The procedure is particularly useful if G is residually nilpotent, i.e. if $\bigcap_{i \in \mathbb{N}} \gamma_i(G) = 1$. Form the direct sum $L = \bigoplus_{i=1}^{\infty} L_i$ of the abelian groups $L_i := \gamma_i(G)/\gamma_{i+1}(G)$. Then commutation in G induces a natural binary operation $[\cdot, \cdot]_{\text{Lie}}$ on L: it is defined on homogeneous elements $x\gamma_{i+1}(G) \in L_i$ and $y\gamma_{j+1}(G) \in L_j$ by

$$[x\gamma_{i+1}(G), y\gamma_{j+1}(G)]_{\text{Lie}} := [x, y]\gamma_{i+j+1}(G) \in L_{i+j},$$

and can be uniquely extended to yield a bi-additive operation on L. As $[y, x] = [x, y]^{-1}$ for all $x, y \in G$, this binary operation on L is skew-symmetric. Moreover, the Hall–Witt identity can be used to show that $[\cdot, \cdot]_{\text{Lie}}$ satisfies the Jacobi identity. Thus $L = \bigoplus_{i=1}^{\infty} L_i$ obtains the structure of a Lie ring. This Lie ring is graded in the sense that $[L_i, L_j] \subseteq L_{i+j}$ for all indices i, j. If all the homogeneous components L_i happen to have exponent p, we can regard L even as a Lie algebra over $\mathbb{F}_p$. An example of this construction is described in Exercise 4.3.

A more sophisticated way of constructing a Lie ring from a group is based on the so-called Hausdorff formula, which can be regarded as the centre piece of Lie theory.[2] This formula will be discussed in more detail in Section 7.4. Stated briefly, the Hausdorff formula gives an expression for the formal power series

$$\Phi(X, Y) = \log(\exp(X) \cdot \exp(Y)) \in \mathbb{Q}\langle\!\langle X, Y \rangle\!\rangle$$

in non-commuting indeterminates X, Y. Here

$$\log(1 + X) = \sum_{n=1}^{\infty} (-1)^{n-1} X^n / n \quad \text{and} \quad \exp(X) = \sum_{n=0}^{\infty} X^n / n!$$

---

[2]Often the Hausdorff formula is more decoratively referred to as the Baker–Campbell–Hausdorff formula. We apply Bourbaki's short terminology.

denote the usual formal power series. The Hausdorff formula enables one to translate between a Lie ring and a group via the logarithm and exponential functions. Classical and important instances of this procedure are the correspondences of Mal'cev and Lazard; see [29, Ch. 10]. These can be employed, in particular, to study finitely generated torsion-free nilpotent groups and finite p-groups of nilpotency class less than p. Without specifying further details at this point, we formulate:

Theorem 2.1 (Lazard's correspondence). *The Hausdorff formula and its inverse set up a correspondence between:*

- *finite p-groups of nilpotency class less than p and*

- *nilpotent Lie rings of class less than p whose additive group is a finite p-group.*

The correspondence preserves such invariants as the orders and the nilpotency classes of the objects involved.

We give a simple illustration of Lazard's correspondence by describing the isomorphism classes of groups of order p^3 for odd primes p. Writing C_n to denote a cyclic group of order n, there are (up to isomorphism) precisely three abelian groups of order p^3, namely

$$G_1 = C_p \times C_p \times C_p, \quad G_2 = C_{p^2} \times C_p, \quad G_3 = C_{p^3}.$$

We claim that in addition to these there are (up to isomorphism) precisely two non-abelian groups of order p^3. Since the nilpotency class of a group of order p^3 is at most 2, by Lazard's correspondence it suffices to show that there are (up to isomorphism) precisely two nilpotent Lie rings of class 2 and order p^3. Clearly, the underlying additive group of such a Lie ring L cannot be cyclic. Moreover, the commutator Lie subring $[L, L]$ has to coincide with the centre $Z(L)$. From this one shows that each of the two non-cyclic abelian groups of order p^3 supports essentially one nilpotent Lie ring structure. The two resulting Lie rings and their corresponding groups can be realised in terms of matrices over $\mathbb{F}_p$ and $\mathbb{Z}/p^2\mathbb{Z}$, respectively; see Exercise 4.2. Representatives for the two isomorphism classes of non-abelian groups of order p^3 are also given by the following group presentations

$$G_4 = \langle x, y, z \mid x^p = y^p = z^p = 1, z = [x, y], [z, x] = [z, y] = 1 \rangle,$$
$$G_5 = \langle x, y \mid x^{p^2} = y^p = 1, [x, y] = x^p \rangle.$$

In 'real life', Lazard's correspondence forms the starting point for the rather more sophisticated enumeration of finite p-groups of higher order, p^7 say. In [48], O'Brien and Vaughan-Lee show that for $p > 5$ the number of groups of order p^7 is precisely $3p^5 + 12p^4 + 44p^3 + 170p^2 + 707p + 2455 + (4p^2 + 44p + 291)\gcd(p - 1, 3) + (p^2 + 19p + 135)\gcd(p - 1, 4) + (3p + 31)\gcd(p - 1, 5) + 4\gcd(p - 1, 7) + 5\gcd(p - 1, 8) + \gcd(p - 1, 9)$, if I copied everything correctly. In particular, this substantiates for $k = 7$ Higman's famous PORC conjecture, which states that the precise number of groups of order p^k is given by a polynomial in p, depending on k and the residue class of p with respect to a suitable modulus $n(k)$; see [4, §21.1]. PORC stands for 'polynomial on residue classes'.

Note that the graded Lie rings associated to G_4 and G_5 with respect to their lower central series coincide. This illustrates that the first and simpler method, which we presented above, incurs a loss of information. The Lie rings, which can be associated via the Hausdorff formula to suitable pro-p groups, are Lie lattices over the p-adic integers. It turns out that they do in fact determine the pro-p groups completely; see Section 8.4.

2.5 Absolute values

The traditional way to describe the size of a rational number is through the use of absolute values. An *absolute value* on a field K is a real-valued function $|\cdot| : K \to [0, \infty)$, which is non-degenerate, multiplicative and satisfies the triangle inequality; this means that for all $x, y \in K$ we have:

(1) $|x| = 0$ if and only if $x = 0$,

(2) $|xy| = |x| \cdot |y|$,

(3) $|x + y| \le |x| + |y|$.

The absolute value is *trivial* if $|x| = 1$ for all $x \ne 0$. For our purposes, the absolute value is said to be either *non-archimedean* or *archimedean* according to whether or not it satisfies the ultrametric triangle inequality:

(3′) $|x + y| \le \max\{|x|, |y|\}$.

The ordinary absolute value on $\mathbb{R}$, which is given by $|x|_\infty = \max\{x, -x\}$, restricts to an archimedean absolute value on $\mathbb{Q}$.

In addition, there is an infinite family of non-archimedean absolute values on $\mathbb{Q}$, one for each prime p. Each rational number $x \ne 0$ can be written uniquely in the form

$$x = p^n \cdot \frac{a}{b}, \quad \text{where } n, a, b \in \mathbb{Z} \text{ with } b > 0, \ \gcd(a, b) = 1, \ p \nmid ab.$$

We put

$$v_p(x) := n \quad \text{and} \quad |x|_p := p^{-n}.$$

Setting $v_p(0) := \infty$ and $|0|_p := 0$, we obtain the *p-adic absolute value* $|\cdot|_p$ on $\mathbb{Q}$. Intuitively, x is p-adically small if it is divisible by a large power of p. The map $v_p : \mathbb{Q} \to \mathbb{Z} \cup \{\infty\}$, which captures the same information as $|\cdot|_p$, is called the *p-adic valuation* on $\mathbb{Q}$.

A theorem of Ostrowski states that, up to a suitable equivalence, the ordinary absolute value and the p-adic absolute values exhaust all possible non-trivial absolute values on $\mathbb{Q}$; see [16, §3.1]. They are linked by the curious adelic formula

$$|x|_\infty \prod_p |x|_p = 1 \quad \text{for all } x \in \mathbb{Q},$$

which is intimately linked with the Fundamental Theorem of Arithmetic; see Exercise 4.4.

2.6 p-adic numbers

The field $\mathbb{R}$ of real numbers can be regarded as the completion of $\mathbb{Q}$ with respect to the metric $d_\infty(x,y) = |x - y|_\infty$ induced by the ordinary archimedean absolute value. Formally, one can construct $\mathbb{R}$ by adjoining all the missing limits of Cauchy sequences in $\mathbb{Q}$ with respect to d_∞. Every element $\alpha \in \mathbb{R}$ is the limit of a Cauchy sequence, $\alpha = \lim_{n \to \infty} x_n$ with $x_n \in \mathbb{Q}$, and the absolute value extends to $\mathbb{R}$ via $|\alpha|_\infty = \lim_{n \to \infty} |x_n|_\infty$. By a similar procedure, the ring operations, addition and multiplication, extend from $\mathbb{Q}$ to $\mathbb{R}$, and one obtains again a field with an absolute value. In using the common decimal notation, we tend to think of a real number α as the limit of a particular Cauchy sequence of the form

$$\alpha = \lim_{n \to \infty} x_n, \quad x_n = \lfloor \alpha \rfloor + \sum_{k=1}^{n} a_k \cdot 10^{-k},$$

where $\lfloor \alpha \rfloor$ denotes the integral part of α and the 'digits' a_k are taken from the set $\{0, 1, \ldots, 9\}$. We remark that base 10 is chosen by convention, not for any intrinsic mathematical reason.

Similarly, we can form for each prime p the completion $\mathbb{Q}_p$ of $\mathbb{Q}$ with respect to the metric $d_p(x,y) = |x - y|_p$ induced by the p-adic absolute value. In this case, one has to adjoin all the missing limits of (equivalence classes of) Cauchy sequences with respect to d_p. Every element $\alpha \in \mathbb{Q}_p$ is the limit of a Cauchy sequence, $\alpha = \lim_{n \to \infty} x_n$ with $x_n \in \mathbb{Q}$, and the absolute value is extended to $\mathbb{Q}_p$ by setting $|\alpha|_p = \lim_{n \to \infty} |x_n|_p$. Similarly, one extends to $\mathbb{Q}_p$ the valuation map v_p and the ring operations, addition and multiplication. One then checks that $\mathbb{Q}_p$ is again a field with absolute value $|\cdot|_p$. The elements of $\mathbb{Q}_p$ are called *p-adic numbers*. For details of this construction, we refer to [16, §3.2].

A convenient notation for explicit computations with p-adic numbers is the following. Every $\alpha \in \mathbb{Q}_p$ can be written uniquely as a series

$$\alpha = \sum_{k \in \mathbb{Z}} a_k p^k = \sum_{k=v_p(\alpha)}^{\infty} a_k p^k,$$

where the coefficients a_k are taken from the set $R_p := \{0, 1, \ldots, p - 1\}$, $a_k = 0$ for $k < v_p(\alpha)$, and $a_{v_p(\alpha)} \neq 0$ if $\alpha \neq 0$. We remark that instead of R_p we could use any set of representatives for $\mathbb{Z}$ modulo $p\mathbb{Z}$. Particularly useful are the so-called Teichmüller representatives consisting of 0 and the $p - 1$ th roots of unity; see [16, §4.5].

A central feature of the real numbers $\mathbb{R}$ is that the field operations are continuous with respect to (the topology underlying) the metric associated to the absolute value $|\cdot|_\infty$. Section 3 contains a summary of basic notions in topology, which we assume. As a topological space, $\mathbb{R}$ is Hausdorff, locally compact and connected. Approximate computations in $\mathbb{R}$ can be performed by truncating the decimal representations of the numbers involved and verifying that errors do not pile up too much – the last bit can actually be quite tricky. Indeed, bounding error terms is one of the main themes in analysis.

In a similar way, the p-adic absolute value induces a metric and hence a topology on $\mathbb{Q}_p$. It is an inherent feature of the completion process that the field operations are continuous. As a topological space, $\mathbb{Q}_p$ is Hausdorff, locally compact and totally disconnected. In fact, if we regard R_p as a finite discrete space and endow $\prod_{k\in\mathbb{Z}} R_p$ with the product topology, then the coordinate map

$$\mathbb{Q}_p \to \prod_{k\in\mathbb{Z}} R_p, \quad \alpha = \sum_{k\in\mathbb{Z}} a_k p^k = \sum_{k=v_p(\alpha)}^{\infty} a_k p^k \mapsto (a_k)$$

is a homeomorphism from $\mathbb{Q}_p$ onto the open subspace

$$\left\{ (a_k)_{k\in\mathbb{Z}} \in \prod_{k\in\mathbb{Z}} R_p \mid \exists n\, \forall k < n : a_k = 0 \right\}.$$

Approximate computations in $\mathbb{Q}_p$ can be performed by truncating the standard representations of the numbers involved; the ultrametric triangle inequality guarantees that errors will not accumulate as in classical analysis; see Exercise 4.4.

A *local field* is a field K, equipped with a non-trivial, non-archimedean absolute value, such that K is locally compact with respect to the induced topology. If K is a finite extension of $\mathbb{Q}_p$, then the p-adic absolute value $|\cdot|_p$ extends uniquely to an absolute value on K, and K becomes a local field. Conversely, it can be shown that every local field of characteristic 0 arises in this manner; cf. [58, I.§3].

2.7 p-adic integers

Finally, we describe a most notable difference between the archimedean field $\mathbb{R}$ and its counterparts, the non-archimedean p-adic fields $\mathbb{Q}_p$. The ultrametric triangle inequality implies that the open compact set

$$\mathbb{Z}_p := \{\alpha \in \mathbb{Q}_p \mid |\alpha|_p \le 1\} = \left\{ \sum_{k=0}^{\infty} a_k p^k \mid a_k \in R_p \right\}$$

forms a subring of $\mathbb{Q}_p$. It is the topological closure of the ordinary integers $\mathbb{Z}$ in $\mathbb{Q}_p$, and its elements are called *p-adic integers*.

The structure of the ring of p-adic integers is quite simple. A short computation reveals that its group of units is given by

$$\mathbb{Z}_p^* = \{\alpha \in \mathbb{Q}_p \mid |\alpha|_p = 1\} = \left\{ \sum_{k=0}^{\infty} a_k p^k \mid a_k \in R_p \text{ and } a_0 \ne 0 \right\}.$$

Moreover, the ideals of $\mathbb{Z}_p$ are principal and of the form $p^n\mathbb{Z}_p$ or $\{0\}$: they line up neatly in a descending chain

$$\mathbb{Z}_p \supset p\mathbb{Z}_p \supset p^2\mathbb{Z}_p \supset \ldots \supset \{0\}.$$

The proper quotient rings of $\mathbb{Z}_p$ are the familiar finite rings $\mathbb{Z}_p/p^n\mathbb{Z}_p \cong \mathbb{Z}/p^n\mathbb{Z}$. In Section 5.3, we describe how $\mathbb{Z}_p$ can be regarded as an inverse limit of these finite quotients. Intuitively, one should think of performing ring operations in $\mathbb{Z}_p$ as follows: do the operations in the ring $\mathbb{Z}_p/p^n\mathbb{Z}_p$, then let n tend to infinity.

2.8   Preview: *p*-adic analytic pro-*p* groups

As yet we have not even defined what we mean by a pro-*p* group; the concept will be introduced in Section 5.6. Nevertheless, skipping the theory that lies in between, we can already formulate a precise and hands-on description of the family of compact *p*-adic Lie groups and *p*-adic analytic pro-*p* groups, in particular.

A *topological group* is a group G which is also a topological space such that the group operations are continuous, i.e. such that the map $G \times G \to G$, $(g,h) \mapsto g^{-1}h$ is continuous. Let $d \in \mathbb{N}$, and consider the group $\mathrm{GL}_d(\mathbb{Z}_p)$ of all invertible $d \times d$ matrices over the ring $\mathbb{Z}_p$ of *p*-adic integers. The set $\mathrm{Mat}_d(\mathbb{Z}_p)$ of all $d \times d$ matrices carries a natural *p*-adic topology, namely the product topology induced from the *p*-adic topology on $\mathbb{Z}_p$. Matrix multiplication is easily seen to be continuous, and so is the process of forming the inverse of an invertible matrix. Hence $\mathrm{GL}_d(\mathbb{Z}_p)$, equipped with the subspace topology, becomes a topological group. We can now state:

Theorem 2.2 (Lazard's characterisation of compact *p*-adic Lie groups). *A compact topological group admits a p-adic analytic structure if and only if it is isomorphic to a closed subgroup of* $\mathrm{GL}_d(\mathbb{Z}_p)$ *for a suitable degree d.*

In fact, in his seminal paper [39] Lazard established a whole theory of *p*-adic analytic groups with much wider consequences. One of his key results is that the analytic structure of a *p*-adic analytic group is determined entirely by its topological group structure. This can be regarded as a positive solution to Hilbert's fifth problem for *p*-adic Lie groups.

As we will see, $\mathrm{GL}_d(\mathbb{Z}_p)$ is virtually a pro-*p* group. This means that $\mathrm{GL}_d(\mathbb{Z}_p)$ contains a subgroup of finite index which is a pro-*p* group. Theorem 2.2 implies that every compact *p*-adic Lie group is virtually a pro-*p* group.

We conclude this section with a concrete reformulation of Theorem 2.2 which applies more directly to pro-*p* groups. There is a natural ring homomorphism from $\mathbb{Z}_p$ onto the finite prime field $\mathbb{F}_p$. As $\mathbb{Z}_p^* = \mathbb{Z}_p \backslash p\mathbb{Z}_p$, this induces a surjective group homomorphism $\eta : \mathrm{GL}_d(\mathbb{Z}_p) \to \mathrm{GL}_d(\mathbb{F}_p)$. The kernel of η is the first congruence subgroup $\mathrm{GL}_d^1(\mathbb{Z}_p) = \{g \in \mathrm{GL}_d(\mathbb{Z}_p) \mid g \equiv 1 \pmod{p}\}$ of $\mathrm{GL}_d(\mathbb{Z}_p)$. The preimage under η of any Sylow *p*-subgroup of the finite group $\mathrm{GL}_d(\mathbb{F}_p)$ constitutes a Sylow pro-*p* subgroup of $\mathrm{GL}_d(\mathbb{Z}_p)$; cf. Section 5.6 and Exercise 6.4. One particular Sylow *p*-subgroup of $\mathrm{GL}_d(\mathbb{F}_p)$ is the group of upper uni-triangular matrices; according to the Sylow theorems, all other Sylow p-subgroups of $\mathrm{GL}_d(\mathbb{F}_p)$ are conjugate to this one.

Corollary 2.3. *A p-adic analytic pro-p group is a topological group which is isomorphic to a closed subgroup of a Sylow pro-p subgroup of $\mathrm{GL}_d(\mathbb{Z}_p)$ for a suitable degree d.*

Suggestions for further reading

Philip Hall's Edmonton notes [17] form a classical text on nilpotent groups. Introductory accounts of the theory of finite p-groups include [40], by Leedham-Green and McKay, who cover the classification of finite p-groups by coclass, and [29], by Khukhro, who considers automorphisms of finite p-groups. The latter covers in detail Lie methods, in particular the Mal'cev and Lazard correspondences. Lazard's classical article [38] covers comprehensively the interrelations between groups and associated Lie rings. For a stimulating survey on Lie methods in the group theory, see, for instance, Shalev's chapter in [11]. For topics related to Higman's PORC conjecture, we refer to the accessible monograph [4] by Blackburn, Neumann and Venkataraman. A readable introduction to p-adic numbers, including many instructive exercises and a guide to more advanced books, is given by Gouvêa in [16]. Neukirch's concise chapter in [12] on the p-adic numbers is also recommended.

3 Basic notions and facts from point-set topology

Pro-p groups and, more generally, profinite groups form a particular class of topological groups. To discuss their structure, we require basic notions and facts from point-set topology. For the convenience of the reader, I have listed the relevant prerequisites below. Suitable references for general topology are, for instance, [5, 28, 59].

A *topological space* $X = (X, \tau)$ is a set X together with a topology, given by a collection τ of *open* subsets of X, satisfying: (i) X and $\varnothing$ are open; (ii) the union of any family of open sets is open; (iii) the intersection of any two open sets is open. The complement in X of any open set is called a *closed* set. It is convenient to use the notation $A \subseteq_{\mathrm{o}} X$ (respectively $A \subseteq_{\mathrm{c}} X$) to indicate that a subset A is open (respectively closed) in X. Every metric space (X, d) with distance function d has an underlying topology: the open sets in this topology are the unions of 'open' balls $\{y \mid d(x, y) < r\}$, where $x \in X$ and $r \in \mathbb{R}$. The *discrete* topology on a set X is the topology in which every subset of X is open.

Let $\varphi : X \to Y$ be a map between topological spaces. The map φ is *continuous* if the preimage of any open set is open, i.e. if $B\varphi^{-1} \subseteq_{\mathrm{o}} X$ for all $B \subseteq_{\mathrm{o}} Y$. The map $\varphi : X \to Y$ is a *homeomorphism* if it is continuous, bijective and admits a continuous inverse.

Let X be a topological space. The *subspace topology* on a subset $Y \subseteq X$ is defined by declaring all intersections $Y \cap A$ with $A \subseteq_o X$ to be open. This is the smallest topology which renders the natural inclusion $Y \to X$ continuous. The *quotient topology* on $Y := X/\sim$ with respect to an equivalence relation $\sim$ is defined by declaring a subset $B \subseteq Y$ open if its preimage in X under the natural projection is open. This is the smallest topology which renders the projection $X \to Y$ continuous.

If X_i, $i \in I$, is a family of topological spaces, then the *product topology* on the Cartesian product $X := \prod_{i \in I} X_i$ is defined by declaring a subset of X open if it is the union of basic open sets of the form $\prod_{i \in I} U_i$, where $U_i = X_i$ for almost all $i \in I$ and $U_i \subseteq_o X_i$ for all $i \in I$. It is the smallest topology such that all canonical projections $X \to X_i$, $i \in I$, are continuous.

Let X be a topological space. The *closure* $\mathrm{cl}(A)$ of a subset $A \subseteq X$ is the intersection of all closed sets containing A; it constitutes the smallest closed set containing A. A subset $A \subseteq X$ is *dense* in X if its closure is equal to X. If $x \in A \subseteq_o X$, then A is called an *open neighbourhood* of x. The space X is *Hausdorff* if any two distinct points have disjoint open neighbourhoods. The space X is *compact* if any covering $X = \bigcup\{U_i \mid i \in I\}$ of X by open subsets $U_i \subseteq_o X$ admits a finite subcovering $X = \bigcup\{U_i \mid i \in J\}$, $J \subseteq I$ with $|J| < \infty$.[3] Equivalently, X is compact if for any non-empty family C_i, $i \in I$, of closed subsets of X having the finite intersection property

$$\bigcap\{C_i \mid i \in J\} \neq \varnothing \quad \text{for all } J \subseteq I \text{ with } 1 \leq |J| < \infty$$

one has $\bigcap\{C_i \mid i \in I\} \neq \varnothing$. The continuous image of a compact space is compact. A theorem of Tychonoff states that the product of any family of compact spaces is compact.

The space X is *locally compact* if every point $x \in X$ has a local base of compact neighbourhoods, i.e. if for every open neighbourhood U of x there exist $V \subseteq_o X$ and a compact subset $C \subseteq X$ such that $x \in V \subseteq C \subseteq U$. Every compact Hausdorff space is locally compact; see Exercise 4.5.

The space X is *connected* if it cannot be partitioned into two proper open subsets, i.e. if for all $A \subseteq_o X$ with $X \backslash A \subseteq_o X$ one has $A = \varnothing$ or $A = X$. Equivalently, X is connected if every continuous map from X into the discrete space $\{0, 1\}$ is constant. The continuous image of a connected space is connected. The maximal non-empty connected subsets of X are called *connected components* of X. The connected components of X are closed and they partition X. The space X is *totally disconnected* if all its connected components are one-point sets. We note that the product of any family of totally disconnected spaces is totally disconnected; cf. Exercise 4.5. A *path* from x to y in the space X is a continuous map φ from the unit interval $[0, 1]$ into X with $\varphi(0) = x$ and $\varphi(1) = y$. The space X is *path-connected* if any two points of X can be joined by a path. Every path-connected space is connected.

---

[3]The original notion 'kompakt' due to Hausdorff – and also adopted by Bourbaki – is reserved for spaces which are compact, in the given sense, and Hausdorff.

4   First series of exercises

This first series of exercises is also intended to serve as a bridge toward topics which will be covered in later sections. The reader is not necessarily expected to solve all parts of all exercises at the first go.

Exercise 4.1 (Finite p-groups).
(a) Let N be a non-trivial normal subgroup of a non-trivial finite p-group G. Show that $Z(G) \cap N \neq 1$. Conclude that $Z(G) \neq 1$ and that G is nilpotent.
(b) Prove that the nilpotency class of a group of order p^n is at most $n-1$. Construct a group of order p^p and nilpotency class $p-1$ along the following lines. Let V be a p-dimensional vector space over $\mathbb{F}_p$ with basis $e_1, \ldots, e_p$. Consider the linear map $\alpha : V \to V$, given by $e_i^\alpha = e_{i+1}$ for $i \in \{1, \ldots, p-1\}$ and $e_p^\alpha = e_1$. Observe that the 1-dimensional subspace U of V, which is spanned by $e_1 + \ldots + e_p$, is invariant under α. Consider the semidirect product of V/U by $\langle \alpha \rangle$.
Remark: Groups of order p^n and nilpotency class $n-1$ are said to be of *maximal class*; they are also known as groups of coclass 1. The semidirect product of V by $\langle \alpha \rangle$ is isomorphic to the wreath product $C_p \wr C_p$.
(c) Let $n \in \mathbb{N}$ and write $n = n_0 + n_1 p + \ldots + n_r p^r$ with $0 \leq n_i < p$ for $i \in \{0, \ldots, r\}$. Determine $v_p(n!)$, i.e. the exponent of the highest p-power dividing $n!$, in terms of the numbers n_i. (*Hint:* First describe $v_p(n!)$ in terms of the numbers $\lfloor n/p^i \rfloor$.)
The *wreath product* $C_p \wr H$ of the cyclic group C_p with a finite permutation group H of degree n is the semidirect product of the base group C_p^n by H, with H acting by coordinate permutations. The group $C_p \wr H$ has a natural permutation action of degree np. Let $k \in \mathbb{N}$ and observe that the iterated wreath product $W_k := C_p \wr (C_p \wr \ldots \wr C_p)$ of k cyclic groups of order p acts naturally on the finite p-regular rooted tree of length k, depicted in Figure I.1 for $p = 3$ and $k = 2$.

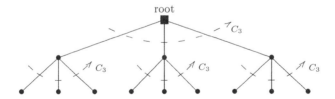

Figure I.1: The wreath product $C_3 \wr C_3 \cong C_3 \ltimes (C_3 \times C_3 \times C_3)$ acts naturally on the rooted 3-regular tree of length 2. In this action the root vertex is fixed, and the action is recorded faithfully on the bottom layer of nine vertices. This describes an embedding of $C_3 \wr C_3$ into the symmetric group $\mathrm{Sym}(9)$.

By computing the order of W_k, prove that the Sylow p-subgroup of the symmetric group $\mathrm{Sym}(p^k)$ is isomorphic to W_k. Conclude that every finite p-group of order at most p^k embeds into W_k.
Can you guess the structure of a Sylow p-subgroup of the symmetric group $\mathrm{Sym}(n)$?

Exercise 4.2 (Groups and Lie rings of order p^3).
(a) Determine up to isomorphism all groups of order p, p^2 and p^3.
(b) Determine up to isomorphism all Lie rings of order p and p^2. Can you find
a Lie ring L of order p^3 which is perfect, i.e. which satisfies $L = [L, L]$?
(c) Show that the groups G_4 and G_5 of order p^3, which are defined in Section 2.4
by presentations, are isomorphic to the subgroups

$$\widetilde{G}_4 = \left\{ \begin{pmatrix} 1 & a & c \\ 0 & 1 & b \\ 0 & 0 & 1 \end{pmatrix} \mid a, b, c \in \mathbb{F}_p \right\}, \quad \widetilde{G}_5 = \left\{ \begin{pmatrix} 1+pa & b \\ 0 & 1 \end{pmatrix} \mid a, b \in \mathbb{Z}/p^2\mathbb{Z} \right\}$$

of $\mathrm{GL}_3(\mathbb{F}_p)$ and $\mathrm{GL}_2(\mathbb{Z}/p^2\mathbb{Z})$, respectively.
Remark: The group $\widetilde{G}_4$ is the finite Heisenberg group over the field $\mathbb{F}_p$.
(d) Show that the sets

$$\widetilde{L}_4 = \left\{ \begin{pmatrix} 0 & a & c \\ 0 & 0 & b \\ 0 & 0 & 0 \end{pmatrix} \mid a, b, c \in \mathbb{F}_p \right\}, \quad \widetilde{L}_5 = \left\{ \begin{pmatrix} pa & b \\ 0 & 0 \end{pmatrix} \mid a, b \in \mathbb{Z}/p^2\mathbb{Z} \right\}$$

form nilpotent Lie subrings of class 2 in $\mathfrak{gl}_3(\mathbb{F}_p)$ and $\mathfrak{gl}_2(\mathbb{Z}/p^2\mathbb{Z})$, respectively.
(e) Let $i \in \{4, 5\}$, and suppose that $p > 2$. Can you see how $\widetilde{L}_i$ and $\widetilde{G}_i$ are related
to one another via the truncated exponential function $X \mapsto 1 + X + X^2/2$ and
the truncated logarithm function $1 + X \mapsto X - X^2/2$?
Set up a bijection $\varphi : \widetilde{L}_i \to \widetilde{G}_i$ based on the truncated exponential function and
work out a Lie expression in $x, y \in \widetilde{L}_i$ for the element $(x^\varphi \cdot y^\varphi)^{\varphi^{-1}}$.
Remark: This will constitute a first approximation to the Hausdorff formula; cf.
Section 7.4.

Exercise 4.3 (The lower central series of the Nottingham group).
The Nottingham group over $\mathbb{F}_p$ is the group G of formal power series

$$\mathbf{a} = t\left(1 + \sum_{i=1}^{\infty} a_i t^i\right) = t + a_1 t^2 + a_2 t^3 + \dots \quad \in t + t^2 \mathbb{F}_p[\![t]\!]$$

with composition given by substitution: the product of $\mathbf{a}, \mathbf{b} \in G$ is defined as

$$\mathbf{a} \circ \mathbf{b} := \mathbf{a}(\mathbf{b}(t)) = t + (a_1 + b_1)t^2 + (a_2 + 2a_1 b_1 + b_2)t^3 +$$
$$(a_3 + 3a_2 b_1 + a_1 b_1^2 + 2a_1 b_2 + b_3)t^4 + \dots$$

(a) Convince yourself that every element of G has an inverse with respect to the
prescribed composition.
(b) Consider the elements $\mathbf{e}_i := t + t^{i+1} \in G$, $i \in \mathbb{N}$. Verify that $\mathbf{e}_i^\lambda \equiv t + \lambda t^{i+1}$
modulo t^{i+2} for $i \in \mathbb{N}$ and $\lambda \in \mathbb{Z}$. Prove that $\mathbf{e}_i \circ \mathbf{e}_j \equiv \mathbf{e}_j \circ \mathbf{e}_i$ modulo t^{i+j+1} and
deduce that $[\mathbf{e}_i, \mathbf{e}_j] \equiv \mathbf{e}_{i+j}^{i-j}$ modulo t^{i+j+2} for all $i, j \in \mathbb{N}$.
(c) Show that for every $n \in \mathbb{N}$ the set $G_n := \{\mathbf{a} \in G \mid \mathbf{a} \equiv t \mod t^{n+1}\}$ forms
a normal subgroup of index p^{n-1} in G. Show that $G_n = \langle \mathbf{e}_n \rangle G_{n+1}$ for $n \in \mathbb{N}$.

(d) Let $n \in \mathbb{N}$, and put $\Gamma_n := G/G_n$. Write e_i for the image of $\mathbf{e}_i$ in Γ_n. Show that every element $g \in \Gamma_n$ can be written uniquely in the form $g = e_1^{\lambda_1} e_2^{\lambda_2} \cdots e_{n-1}^{\lambda_{n-1}}$ with exponents $\lambda_i \in \{0, 1, \ldots, p-1\}$.
(e) Let $n \in \mathbb{N}$, and suppose that $p > 2$. Determine the commutator subgroup $[\Gamma_n, \Gamma_n]$ of Γ_n and show that it coincides with the Frattini subgroup. Conclude that e_1 and e_2 form a minimal generating pair for Γ_n.
(f) Suppose that $p > 2$. Work out the lower central series for Γ_{p+1} and determine the graded Lie ring associated to this group with respect to its lower central series, as described in Section 2.4.
(g) Suppose that $p > 2$. Work out the lower central series of Γ_{p+2}. Can you guess the general pattern of the lower central series of Γ_n as $n \to \infty$? See whether your guess is consistent with the following formula: for $n \geq 3$ the nilpotency class of Γ_n is equal to $(n-2) - \lfloor (n-3)/p \rfloor$.

Exercise 4.4 (Non-archimedean absolute values and p-adic numbers).
(a) Prove that the adelic formula stated in Section 2.5 holds.
(b) Compute the standard representation $\sum_{k=0}^{\infty} a_k 3^k$ of -13 in the ring $\mathbb{Z}_3$ of 3-adic integers. Show that 5 has a multiplicative inverse in the ring $\mathbb{Z}_3$, by displaying the standard representation of such an element. Prove that 11 has no square root in $\mathbb{Z}_3$, but convince yourself that 7 does by computing the first five coefficients of the standard representation of a potential root.
Prove that 2 has no square root in $\mathbb{Q}_2$. Now suppose that $p > 2$. Convince yourself that 2 has a square root in $\mathbb{Q}_p$ if and only if it has one in $\mathbb{Z}_p$. Then show that 2 has a square root in $\mathbb{Z}_p$ if and only if 2 admits a square root modulo p. (*Hint:* Look at the quotient $\mathbb{Z}_p/p\mathbb{Z}_p$ to see that the condition is necessary. Now suppose that $x \in \mathbb{Z}_p$ satisfies $x^2 - 2 \equiv 0$ modulo p. Then $x^2 - 2 = pa$ for a suitable $a \in \mathbb{Z}_p$. Write $\tilde{x} = x + py$ with $y \in \mathbb{Z}_p$ to be specified. Since $2x \in \mathbb{Z}_p^*$, the congruence

$$p(2xy + a) \equiv x^2 + 2pxy + p^2y^2 - 2 = \tilde{x}^2 - 2 \equiv 0 \pmod{p^2}$$

can be solved for $y \in \mathbb{Z}_p$. Now continue inductively to find a Cauchy sequence $x, \tilde{x}, \ldots$ in $\mathbb{Z}_p$ whose limit gives a precise square root of 2.)
Remark: This procedure is a particular instance of Hensel's lemma; see [16, §3.4].
(c) Let $|\cdot|$ be a non-archimedean absolute value on a field K. Prove that for all $x, y \in K$ with $|x| \neq |y|$ the ultrametric triangle inequality specialises to $|x + y| = \max\{|x|, |y|\}$.
Suppose further that K is complete with respect to the metric $d(x, y) = |x - y|$ induced by the absolute value. Conclude that for any sequence $(a_n)_{n \in \mathbb{N}}$ in K the series $\sum_{n=1}^{\infty} a_n$ converges in K if and only if $|a_n| \to 0$ for $n \to \infty$. For which p does the harmonic series $\sum_{n=1}^{\infty} n^{-1}$ converge in $\mathbb{Q}_p$?
(d) Suppose that $p > 2$. Given that $v_p(n!) \leq (n-1)/(p-1)$ for all $n \in \mathbb{N}$, show that the exponential series $\exp(x) = \sum_{n=0}^{\infty} x^n/n!$ converges for all $x \in p\mathbb{Z}_p$. Deduce that the exponential series induces an isomorphism of topological groups from the additive group $p\mathbb{Z}_p$ onto the multiplicative group of one-units $1 + p\mathbb{Z}_p$.

Remark: Clearly, the additive groups $p\mathbb{Z}_p$ and $\mathbb{Z}_p$ are isomorphic. It can be shown that the subgroup $1 + p\mathbb{Z}_p$ of the abelian group $\mathbb{Z}_p^*$ admits a cyclic complement so that $\mathbb{Z}_p^* \cong \mathbb{Z}_p \times C_{p-1}$; see [16, §4.5].

(e) Show that the additive group $\mathbb{Z}_2$ and the multiplicative group $1 + 2\mathbb{Z}_2$ are not isomorphic. Can you mend the situation by considering a subgroup of finite index in $1 + 2\mathbb{Z}_2$ and subsequently determine the structure of $\mathbb{Z}_2^*$?

Exercise 4.5 (Point-set topology and topological groups).

(a) Show that $\mathrm{GL}_2(\mathbb{R})$, with respect to the natural topology, is a locally compact, Hausdorff topological group. Is this group connected? If not, how many connected components does it have? (*Hint:* Think of determinants and canonical forms of matrices.)
Show that $\mathrm{GL}_2(\mathbb{Z})$ is a discrete subgroup of $\mathrm{GL}_2(\mathbb{R})$. Can you give an example of an infinite compact subgroup of $\mathrm{GL}_2(\mathbb{R})$? (*Hint:* Think of rotation matrices.) Does $\mathrm{GL}_2(\mathbb{R})$ admit any open compact subgroups?

(b) Let X, Y be topological spaces. Prove the following assertions from first principles. (i) If X is Hausdorff, then every compact subset of X is closed. (ii) If X is compact, then every closed subset of X is compact. (iii) If X is compact and Y is Hausdorff, then every continuous bijection $f : X \to Y$ is a homeomorphism. (iv) Every compact Hausdorff space is locally compact.

(c) Regard C_p as a topological group, equipped with the discrete topology. Convince yourself that C_p is totally disconnected, compact and Hausdorff. Using Tychonoff's theorem and first principles, deduce that $G := \prod_{k \in \mathbb{Z}} C_p$ is a totally disconnected, compact, Hausdorff topological group. Does G admit a finitely generated dense subgroup?
Let $V = \bigoplus_{k \in \mathbb{Z}} \mathbb{F}_p e_k$ be a vector space over $\mathbb{F}_p$ of countably infinite dimension. Show that the underlying abelian group of the dual space $\check{V} := \mathrm{Hom}_{\mathbb{F}_p}(V, \mathbb{F}_p)$ is isomorphic to G. What is the dimension of the $\mathbb{F}_p$-vector space $\check{V}$? Does G admit a countably generated dense subgroup?

(d) Let G be a topological group. Prove the following assertions from first principles. (i) For each $g \in G$, the maps $x \mapsto xg$, $x \mapsto gx$ and $x \mapsto x^g$ are homeomorphisms of G. (ii) If H is a subgroup of G and H is closed (respectively open), then every coset of H in G is closed (respectively open). (iii) Every open subgroup of G is closed. (iv) If H is a subgroup of G, then its closure $\mathrm{cl}(H)$ is also a subgroup of G. (v) If H is a subgroup of G and H contains a non-empty open subset of G, then H is open in G. (vi) The group G is Hausdorff if and only if $\{1\}$ is a closed subset of G. (*Hint:* To see that the condition is sufficient consider $x, y \in G$ with $x^{-1}y \neq 1$. In order to find disjoint open neighbourhoods of x and y, look at a suitable open neighbourhood of $(1, 1)$ in $G \times G$ which is fully contained in the preimage of $G \backslash \{x^{-1}y\}$ under the continuous map $(g, h) \mapsto g^{-1}h$.) (v) If N is a closed normal subgroup of G and G is Hausdorff, then G/N is Hausdorff with respect to the quotient topology.

(e) Show that $\mathrm{GL}_2(\mathbb{Q}_p)$, viewed as a topological group with respect to the natural topology, is totally disconnected, locally compact and Hausdorff. Prove that $\mathrm{GL}_2(\mathbb{Z}_p)$ is an open compact subgroup of $\mathrm{GL}_2(\mathbb{Q}_p)$.

Is $\mathrm{GL}_2(\mathbb{Z})$ a dense subgroup of $\mathrm{GL}_2(\mathbb{Z}_p)$? (*Hint:* Try to reduce a given matrix to the identity matrix modulo p^n by elementary row and column operations.) If not, determine the closure of $\mathrm{GL}_2(\mathbb{Z})$ in $\mathrm{GL}_2(\mathbb{Z}_p)$.

Suggestions for further reading

Exercise 4.1 relates to groups acting on spherically homogeneous trees; see Grigorchuk's chapter in [11] on branch groups. The book [40], by Leedham-Green and McKay, contains a detailed account of finite p-groups of maximal class and the classification of finite p-groups by coclass. Exercise 4.3 introduces the Nottingham group; see Camina's chapter in [11] for more information on this remarkable group.

5 Powerful groups, profinite groups and pro-p groups

5.1 Powerful finite p-groups

The theory of finite p-groups and, more generally, pro-p groups is very much governed by the interplay between commutators and pth powers. In some sense it is the right mixture of the two concepts that makes p-adic Lie groups work the way they do. An important class of finite p-groups, defined in terms of this interconnection, is the class of powerful finite p-groups, which was introduced by Mann, and developed by him and Lubotzky in the 1980s; see [42, 43] and [10, Ch. 2]. Another more classical class of finite p-groups, which is defined in terms of commutators and pth powers, comprises the regular p-groups, introduced by Hall in the 1930s. A finite p-group G is *regular* if for all $x, y \in G$ one has $(xy)^p \equiv x^p y^p$ modulo $\gamma_2(\langle x, y \rangle)^p$; see [21, 3.§10].

Let G be a finite p-group. We recall from Section 2.2 that G^p denotes the subgroup generated by all pth powers in G. The group G is *powerful* if p is odd and G/G^p is abelian, or if $p = 2$ and G/G^4 is abelian. More generally, a subgroup $N \leq G$ is *powerfully embedded* in G if p is odd and $[N, G] \subseteq N^p$, or $p = 2$ and $[N, G] \subseteq N^4$.

Thus G is powerful if and only if it is powerfully embedded in itself; and if N is powerfully embedded in G, then $N \trianglelefteq G$ and N is powerful. Recall that $\Phi(G)$ denotes the Frattini subgroup of G. When p is odd, G is powerful if and only if $\Phi(G) = G^p$; for $p = 2$ the equation $\Phi(G) = G^2$ always holds. Clearly, every abelian finite p-group is powerful, and one should think of 'powerful' as a generalisation of 'abelian'.

Proposition 5.1. *If G is a finite p-group and N is powerfully embedded in G, then N^p is powerfully embedded in G.*

Sketch of proof for $p > 2$. Let G be a finite p-group and $N \leq G$ with $[N, G] \subseteq N^p$. It suffices to show that $[N^p, G] \subseteq [N, G]^p$. Passing to the quotient $G/[N, G]^p$, if necessary, we may assume that $[N, G]^p = 1$. Since G is nilpotent, we have

$[K, G] \not\leq K$ for every non-trivial normal subgroup $K \lhd G$. Hence we may further assume that $[[N^p, G], G] = 1$. This implies that $[[N, G], G] \subseteq [N^p, G] \subseteq Z(G)$.

Let $x \in N$ and $g \in G$. Then $[[x, g], x^i] \in Z(G)$ for $i \in \{0, \dots, p-1\}$, and

$$\prod_{i=0}^{p-1} [[x, g], x^i] = \prod_{i=0}^{p-1} [[x, g], x]^i = [[x, g], x]^{p(p-1)/2}.$$

Since p is odd and $[N, G]^p = 1$, this shows that

$$\begin{aligned}
[x^p, g] &= [x, g]^{x^{p-1}} \cdot [x, g]^{x^{p-2}} \cdots [x, g] \\
&= [x, g] \, [[x, g], x^{p-1}] \cdot [x, g] \, [[x, g], x^{p-2}] \cdots [x, g] \\
&= [x, g]^p \prod_{i=0}^{p-1} [[x, g], x^i] \\
&= [x, g]^p \, [[x, g], x]^{p(p-1)/2} = 1.
\end{aligned}$$

Hence $[N^p, G] = 1$, as wanted. $\qquad\qquad\qquad\qquad\qquad\qquad\qquad\qquad\qquad\square$

The *lower p-series* of a group G is the descending series

$$G = P_1(G) \geq P_2(G) \geq \dots, \quad \text{where } P_{i+1}(G) = P_i(G)^p [P_i(G), G].$$

A basic property of this sequence is that $[P_i(G), P_j(G)] \subseteq P_{i+j}(G)$ for all $i, j \in \mathbb{N}$. Now suppose that G is a finite p-group. Then $P_2(G) = \Phi(G)$ and, more generally, $P_{i+1}(G) \supseteq \Phi(P_i(G))$ for all i. The lower p-series of a powerful finite p-group behaves rather well.

Proposition 5.2. *Let* $G = \langle a_1, \dots, a_d \rangle$ *be a powerful finite p-group. Writing* $G_i := P_i(G)$ *for $i \in \mathbb{N}$, the following assertions hold:*

(1) G_i *is powerfully embedded in G;*

(2) $G_{i+k} = P_{k+1}(G_i) = G_i^{p^k}$ *for each $k \in \mathbb{N}$, and in particular $G_{i+1} = \Phi(G_i)$;*

(3) $G_i = G^{p^{i-1}} = \{ x^{p^{i-1}} \mid x \in G \} = \langle a_1^{p^{i-1}}, \dots, a_d^{p^{i-1}} \rangle$;

(4) *the map* $x \mapsto x^{p^k}$ *induces a homomorphism from G_i/G_{i+1} onto G_{i+k}/G_{i+k+1} for each $k \in \mathbb{N}$.*

Corollary 5.3. *If* $G = \langle a_1, \dots, a_d \rangle$ *is a powerful finite p-group, then G decomposes as a product of its cyclic subgroups $\langle a_i \rangle$, i.e. $G = \langle a_1 \rangle \cdots \langle a_d \rangle$.*

Sketch of proof for $p > 2$. The assertions of the proposition and the corollary are established by induction, based on Proposition 5.1; see [10, Theorem 2.7]. As examples we give the proofs of parts (1) and (4).

(1) Since $G_1 = G$ is powerful, the group G_1 is powerfully embedded in G. Suppose that $i \geq 2$. By induction, G_{i-1} is powerfully embedded in G. Then

$G_i = G_{i-1}^p [G_{i-1}, G] = G_{i-1}^p$, and Proposition 5.1 shows that G_i is powerfully embedded in G.

(4) Clearly, it suffices to consider the case $k = 1$. The argument above shows that G_i is powerful, $G_{i+1} = P_2(G_i) = G_i^p$ and $G_{i+2} = P_3(G_i)$. Changing notation, we may assume that $i = 1$. Furthermore, passing from G to G/G_3 we may assume that $G_3 = 1$ so that $[G, G] \subseteq G_2 \subseteq Z(G)$ and $[G, G]^p \subseteq G_2^p \subseteq G_3 = 1$. As p is odd, we have for all $x, y \in G$

$$(xy)^p = x^p y^p [x, y]^{-p(p-1)/2} = x^p y^p.$$

Thus the map $x \mapsto x^p$ induces a homomorphism from G/G_2 onto G_2/G_3.   □

For any group G, let $d(G)$ denote the minimal cardinality of a generating set for G. The *rank* of a finite group G is defined to be $\mathrm{rk}(G) := \max\{d(H) \mid H \le G\}$. If G is a finite p-group, then $d(G)$ is simply the dimension of $G/\Phi(G)$ as a vector space over $\mathbb{F}_p$, but there is no comparable general description of the more subtle invariant $\mathrm{rk}(G)$.

Theorem 5.4. *Let G be a powerful finite p-group. Then $\mathrm{rk}(G) = d(G)$, in other words $d(H) \le d(G)$ for all $H \le G$.*

Proof (by induction on $|G|$). Let $H \le G$ and put $d := d(G)$. Write $G_i := P_i(G)$ for $i \in \mathbb{N}$, and put $d_2 := d(G_2)$. Proposition 5.2 shows that G_2 is powerful, hence by induction the group $K := H \cap G_2$ satisfies $d(K) \le d_2$. Put $e := d(HG_2/G_2)$ so that $e \le d$. Our aim is to find $h_1, \ldots, h_e \in H$ and $y_1, \ldots, y_{d-e} \in K$ such that

$$HG_2 = \langle h_1, \ldots, h_e \rangle G_2 \qquad \text{and} \qquad K = \langle h_1^p, \ldots, h_e^p, y_1, \ldots, y_{d-e} \rangle.$$

This will imply $H = \langle h_1, \ldots, h_e, y_1, \ldots, y_{d-e} \rangle$ and $d(H) \le d$, as wanted.

According to Proposition 5.2, the map $x \mapsto x^p$ induces a homomorphism π from G/G_2 onto G_2/G_3. Both groups are elementary p-groups, so we may regard them as vector spaces over $\mathbb{F}_p$. Basic linear algebra allows us to bound the dimension of the image $(HG_2/G_2)\pi$ over $\mathbb{F}_p$

$$\begin{aligned}
\dim((HG_2/G_2)\pi) &= \dim(HG_2/G_2) - \dim(\ker \pi \cap HG_2/G_2) \\
&\ge \dim(HG_2/G_2) - \dim(\ker \pi) \\
&= \dim(HG_2/G_2) - (\dim(G/G_2) - \dim(G_2/G_3)) \\
&= e - (d - d_2) \\
&= d_2 - (d - e).
\end{aligned}$$

Let $h_1, \ldots, h_e \in H$ such that $HG_2 = \langle h_1, \ldots, h_e \rangle G_2$. Observe that $\Phi(K) \subseteq \Phi(G_2) = G_3$. Hence the subspace of $K/\Phi(K)$ spanned by the cosets of $h_1^p, \ldots, h_e^p$ has dimension at least $\dim((HG_2/G_2)\pi) \ge d_2 - (d - e)$. Since $\dim(K/\Phi(K)) = d(K) \le d_2$, we find $d - e$ elements $y_1, \ldots, y_{d-e} \in K$ such that

$$K = \langle h_1^p, \ldots, h_e^p, y_1, \ldots, y_{d-e} \rangle \Phi(K) = \langle h_1^p, \ldots, h_e^p, y_1, \ldots, y_{d-e} \rangle.$$

□

The naive converse of the theorem is false, but a more complex statement is true: every finite p-group admits a powerful normal subgroup of index bounded by a function of $\mathrm{rk}(G)$.

Theorem 5.5. *Let G be a non-trivial finite p-group of rank $r := \mathrm{rk}(G)$, and write $\lambda(r) := \lceil \log_2(r) \rceil$ if p is odd, $\lambda(r) := \lceil \log_2(r) \rceil + 1$ if $p = 2$. Then G admits a powerful characteristic subgroup of index at most $p^{r\lambda(r)}$.*

This result, which is [10, Theorem 2.13], can be seen as an invitation into the world of pro-p groups. Indeed, it can be translated to characterise pro-p groups of finite rank as virtually powerful pro-p groups; see Section 5.7. Pro-p groups are special kinds of profinite groups, and we shall not delay their introduction any longer.

5.2   Profinite groups as Galois groups

The fundamental theorem of Galois theory sets up a correspondence between the intermediate fields of a finite Galois extension $L|K$ and the subgroups of the associated Galois group $G(L|K)$. In fact, it generalises to infinite Galois extensions, with an interesting twist.

Consider a general Galois extension $L|K$, i.e. a separable splitting field L for a (possibly infinite) family of polynomials over a ground field K. Then L is the union $L = \bigcup \{L_i \mid i \in I\}$ of its finite Galois subextensions $L_i|K$. The set $\{L_i \mid i \in I\}$ is partially ordered by the inclusion relation. It has the property that for all L_i, L_j there exists L_k such that $L_k \supseteq L_i$ and $L_k \supseteq L_j$; just think in terms of splitting fields. Writing $i \succeq j$ whenever $L_i \supseteq L_j$, the last observation can be stated as follows: for all $i, j \in I$ there exists $k \in I$ such that $k \succeq i$ and $k \succeq j$.

The Galois group $G(L|K)$ is, of course, defined as the group of all automorphisms of L which fix K elementwise. Every automorphism $\alpha \in G(L|K)$ is uniquely determined by its restrictions $\alpha|_{L_i}$, $i \in I$, and the normality of $L|K$ and its subextensions $L_i|K$ guarantees that each of the restriction maps $\varphi_i : G(L|K) \to G(L_i|K)$ is onto. Clearly, there is a certain compatibility condition that the restrictions of α to the various L_i satisfy, namely $(\alpha|_{L_i})|_{L_j} = \alpha|_{L_j}$ whenever $L_i \supseteq L_j$. Writing $\varphi_{ij} : G(L_i|K) \to G(L_j|K)$ for the natural restriction map whenever $L_i \supseteq L_j$, the compatibility condition can be rephrased as $\varphi_i \varphi_{ij} = \varphi_j$ whenever $i \succeq j$. A similar condition in terms of the restrictions alone can be stated as $\varphi_{ij}\varphi_{jk} = \varphi_{ik}$ whenever $i \succeq j \succeq k$.

We are now ready to describe the Galois group $G := G(L|K)$. Writing $G_i := G(L_i|K)$, the coordinate map

$$\varphi : G \to \prod_{i \in I} G_i, \quad g \mapsto (g\varphi_i)_{i \in I}$$

induces an isomorphism from G onto the group

$$G\varphi = \left\{ (g_i)_{i \in I} \in \prod_{i \in I} G_i \mid g_i \varphi_{ij} = g_j \text{ whenever } i \succeq j \right\}.$$

But what happens to the Galois correspondence? It turns out that only certain subgroups of G correspond to intermediate fields of $L|K$. To describe why this is so and which subgroups play a role in the Galois correspondence we equip G with the *Krull topology*. Regarding each of the finite Galois groups G_i as a discrete topological group, the product $\prod_{i \in I} G_i$ naturally becomes a topological group which is totally disconnected, compact and Hausdorff. The subgroup $G\varphi$ is easily seen to be closed, so its isomorphic twin G becomes a totally disconnected, compact, Hausdorff topological group. We have seen how the structure of G is determined by its finite images G_i. In Section 5.3, we will formalise this process to see that G is the inverse (or projective) limit of the inverse system $(G_i; \varphi_{ij})$ of finite groups, and thus G becomes a profinite group.

If M is an intermediate field of the extension $L|K$, then the set G^M of all $g \in G$ which fix M elementwise can be described in terms of the restrictions of automorphisms to (the normal closures of) finite subextensions of M. Indeed, G^M can be written as the intersection of closed open subgroups of G and thus forms a closed subgroup with respect to the Krull topology. As described in detail in [60, Ch. 3], the Galois correspondence for finite Galois extensions then readily generalises to yield:

Theorem 5.6 (Fundamental theorem of Galois theory). *The Galois group G of a (typically infinite) Galois extension $L|K$ is a profinite group with respect to the Krull topology. There is an inclusion-reversing correspondence between the lattice of all closed subgroups of G and the lattice of all intermediate fields of $L|K$.*

It can be shown that every profinite group is isomorphic to the Galois group of a suitable Galois extension. One of the fundamental problems in number theory is to describe the finite extensions of a given local field K, such as $\mathbb{Q}_p$. This is equivalent to understanding the absolute Galois group $G(\overline{K}|K)$, where $\overline{K}$ denotes the separable closure of K. Local class field theory provides a rather explicit and very satisfying description of all abelian extensions, i.e. Galois extensions with abelian Galois groups: the lattice of abelian extensions of the local field K has a precise reflection in the multiplicative group K^* via the norm residue symbol. In 1982 Jannsen and Wingberg gave a description of the full Galois group $G(\overline{K}|K)$ of a local field of characteristic 0 in terms of generators and relations; see [26]. A quite different and far-reaching approach is described by the Langlands conjectures; cf. [8].

5.3 Profinite groups as inverse limits

The description of Galois groups in terms of their finite factors can be formalised as follows. A *directed set* is a partially ordered set $I = (I, \preceq)$ such that for all $i, j \in I$ there exists $k \in I$ such that $k \succeq i$ and $k \succeq j$. An *inverse system* $(G_i; \varphi_{ij})$ of groups (or other mathematical structures such as sets, rings, topological spaces, etc.) over I consists of a family of groups (or sets, ...) G_i, $i \in I$, and homomorphisms (or maps, ...) $\varphi_{ij} : G_i \rightarrow G_j$ whenever $i \succeq j$, satisfying the natural compatibility conditions

$$\varphi_{ii} = \mathrm{id}_{G_i} \quad \text{and} \quad \varphi_{ij}\varphi_{jk} = \varphi_{ik} \quad \text{for all } i, j, k \in I \text{ with } i \succeq j \succeq k.$$

The *inverse limit* of the inverse system $(G_i; \varphi_{ij})$ is the group (or set, ...)

$$\varprojlim G_i := \left\{ (g_i)_{i \in I} \in \prod_{i \in I} G_i \mid g_i \varphi_{ij} = g_j \text{ whenever } i \succeq j \right\}$$

together with the natural coordinate maps $\varphi_i : G \to G_i$. It is the (unique) solution to an appropriate universal problem; see Exercise 6.3. In the special, but important case where $I = \mathbb{N}$ and $\preceq$ is the ordinary order-relation $\leq$, we can think of the inverse limit pictorially as the 'limit object' to a chain of homomorphisms as indicated in Figure I.2.

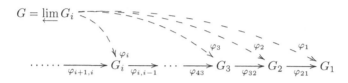

Figure I.2: Pictorial description of the inverse limit G of an inverse system $(G_i; \varphi_{ij})$ of groups (or sets, ...) over the countable index set $I = \mathbb{N}$.

If the G_i are finite groups, we give each of them the discrete topology, and $\prod_{i \in I} G_i$ the product topology. Then $\varprojlim G_i$ with the induced topology becomes a totally disconnected, compact, Hausdorff topological group. Such a group is known as a *profinite group*, which is short for projective limit of finite groups.

Every finite group is a profinite group. As we have seen in the previous subsection, natural examples of infinite profinite groups are given by Galois groups. The simplest such is perhaps the absolute Galois group $G(\overline{\mathbb{F}_q}|\mathbb{F}_q)$ of a finite field $\mathbb{F}_q$. From field theory we know that $\mathbb{F}_q$ has precisely one extension of any given finite degree and that all these extensions are Galois with cyclic Galois group. The corresponding inverse system consists of the cyclic groups $G_n \cong \mathbb{Z}/n\mathbb{Z}$, $n \in \mathbb{N}$, with φ_{mn} given by the natural projections $\mathbb{Z}/m\mathbb{Z} \to \mathbb{Z}/n\mathbb{Z}$ whenever $n \mid m$. The inverse limit of this inverse system is the procyclic group $\hat{\mathbb{Z}} := \varprojlim \mathbb{Z}/n\mathbb{Z}$.

In fact, $\hat{\mathbb{Z}}$ can be regarded as a profinite ring, simply by going through the same construction, considering each $\mathbb{Z}/n\mathbb{Z}$ not simply as a group but as a ring. A similar construction yields the ring $\mathbb{Z}_p$ of p-adic integers. There is an interesting connection between $\hat{\mathbb{Z}}$ and the rings of p-adic integers: $\hat{\mathbb{Z}} = \prod_p \mathbb{Z}_p$ as topological rings; see Exercise 6.1.

5.4　Profinite groups as profinite completions

An impressively fruitful theme in infinite group theory builds upon the following unassuming question: how do the finite images of an infinite group reflect its structure? A group G is *residually finite* if the intersection of all its finite-index subgroups is trivial. Residually finite groups are the groups whose structures one can hope to understand in terms of finite images, and they form quite a large class. For instance, every finitely generated linear group is residually finite, a result that goes back to Mal'cev; see [44, Wind. 7 Prop. 8].

Let Γ be any group. It is convenient to use the notation $H \leq_f \Gamma$ to indicate that H is a subgroup of finite index in Γ. Note that the finite quotients of Γ form a natural inverse system Γ/N, $N \trianglelefteq_f \Gamma$, with φ_{MN} given by the natural projection $\Gamma/M \to \Gamma/N$ whenever $M \subseteq N$. The inverse limit of this inverse system is the *profinite completion* $\hat{\Gamma} := \varprojlim \Gamma/N$ of Γ. There is a natural map from the original group into its profinite completion, namely

$$\vartheta : \Gamma \to \hat{\Gamma}, \quad g \mapsto (gN)_{N \trianglelefteq_f \Gamma}.$$

If Γ is residually finite, then ϑ is injective. Typically, $\Gamma\vartheta$ is strictly contained in $\hat{\Gamma}$, but it always forms a dense subgroup. The notation $\hat{\mathbb{Z}}$ is no coincidence: the procyclic group $\hat{\mathbb{Z}}$ can be regarded as the profinite completion of the infinite cyclic group.

5.5 Profinite groups as topological groups

Profinite groups are topological groups which are totally disconnected, compact and Hausdorff. Indeed, it can be shown that the converse holds; see [54, I.§1] or [60, Cor. 1.2.4]. But for our purposes the following characterisation is perhaps more useful: a profinite group is a compact Hausdorff topological group G such that every open neighbourhood of the neutral element 1 contains an open subgroup. This means that the open subsets of a profinite group G are precisely those sets which can be written as unions of cosets gN of open normal subgroups $N \trianglelefteq_o G$.

Profinite groups are typically quite large, i.e. uncountable, and therefore rather unwieldy as abstract groups. But our interest is mainly focused on closed subgroups. So group-theoretic notions and constructions should be employed with a topological twist. Often it is agreed implicitly that this approach is being taken.

For instance, let G be a profinite group and $X \subseteq G$. Then X is said to *generate* G (topologically) if X generates a dense subgroup of G. Accordingly, G is *finitely generated* (as a topological group) if it admits a finite (topological) generating set. We denote by $d(G)$ the minimal cardinality of a (topological) generating set for G. In order to check whether a given subset X generates a profinite group G, it suffices to show that X generates G modulo every open normal subgroup $N \trianglelefteq_o G$; see Exercise 6.3. Thus one has $d(G) = \sup\{d(G/N) \mid N \trianglelefteq_o G\}$.

The *Frattini subgroup* $\Phi(G)$ of a profinite group G is the intersection of all maximal proper open subgroups of G. Since every open subgroup is closed, it follows that $\Phi(G)$ is a closed subgroup of G. Furthermore, one can show that $X \subseteq G$ generates G if and only if X generates G modulo $\Phi(G)$.

Every open subgroup of a profinite group G has finite index in G. A theorem of Nikolov and Segal, relying on the classification of finite simple groups, states that in a finitely generated profinite group G every finite-index subgroup is open; see [47]. For pro-p groups, this is a result of Serre; see [54, I.§4 Ex. 6] and Exercise 6.5. So in the case of finitely generated profinite groups, the topology is uniquely determined by the algebraic structure of the group.

5.6   Pro-p groups

A *pro-p group* is a topological group which is isomorphic to the inverse limit of finite p-groups. Every group Γ admits a *pro-p completion* $\hat{\Gamma}_p$, which is the pro-p group arising from the inverse system of finite quotients Γ/N where N runs through all normal subgroups of p-power index in Γ.

Let G be a pro-p group. Then every closed subgroup of G is a pro-p group and any quotient of G by a closed normal subgroup is a pro-p group. In particular, the index of any open subgroup of G is a power of p; see Exercise 6.5. The Frattini subgroup of G is equal to the closure of the abstract Frattini subgroup, i.e. $\Phi(G) = \operatorname{cl}(G^p[G,G])$. In particular, one has $d(G) = \dim_{\mathbb{F}_p} G/\Phi(G)$.

The category of pro-p groups is quite large. Our main focus will be on the class of pro-p groups of finite rank (which are the same as p-adic analytic pro-p groups), but we give a variety of examples:

(1) The additive group of p-adic integers $\mathbb{Z}_p$ is the most basic infinite pro-p group. It is an example of a procyclic group and plays a similar role to the infinite cyclic group in abstract group theory; see Exercise 6.1.

(2) The Sylow theorems for finite groups carry over to profinite groups: every pro-p subgroup of a profinite group G is contained in a maximal pro-p subgroup, and any two maximal pro-p subgroups of G are conjugate in G; see Exercise 6.4. Maximal pro-p subgroups of G are called *Sylow pro-p subgroups*.

(3) Matrix groups over $\mathbb{Z}_p$ are virtually pro-p groups; see Exercise 6.2. According to Lazard, they constitute the class of compact p-adic Lie groups; cf. Theorem 2.2. Typical examples of p-adic analytic pro-p groups are the Sylow pro-p subgroups of $\operatorname{GL}_d(\mathbb{Z}_p)$.

(4) Let $d \in \mathbb{N}$ and F a free group on d generators. Then the pro-p completion $\hat{F}_p$, known as a *free pro-p group*, can be seen to be a free object (on d generators) in the category of pro-p groups; see [60, Ch. 5].

(5) The *Nottingham group* over $\mathbb{F}_p$, which was introduced in Exercise 4.3 is a finitely generated pro-p group. It is virtually isomorphic to the automorphism group of a local field of characteristic p. The Nottingham group has remarkable properties, e.g. it can be shown that every finitely generated pro-p group embeds into it as a closed subgroup; see [11, Ch. 6].

Next we return our attention to the concept of powerful groups, which we introduced in Section 5.1.

5.7 Powerful pro-p groups

Let G be a pro-p group. The group G is *powerful* if p is odd and $G/\operatorname{cl}(G^p)$ is abelian, or if $p = 2$ and $G/\operatorname{cl}(G^4)$ is abelian. Equivalently, the pro-p group G is powerful if it is the inverse limit of powerful finite p-groups. More generally, a subgroup $N \leq_c G$ is *powerfully embedded* in G if p is odd and $[N, G] \subseteq \operatorname{cl}(N^p)$, or $p = 2$ and $[N, G] \subseteq \operatorname{cl}(N^4)$. Thus G is powerful if and only if G is powerfully embedded in itself; and if N is powerfully embedded in G, then $N \trianglelefteq_c G$ and N is powerful.

The *lower p-series* of a topological group G is the descending series

$$G = P_1(G) \geq P_2(G) \geq \ldots, \quad \text{where } P_{i+1}(G) = \operatorname{cl}(P_i(G)^p[P_i(G), G]).$$

A basic property of this sequence is that $[P_i(G), P_j(G)] \subseteq P_{i+j}(G)$ for all $i, j \in \mathbb{N}$. Proposition 5.2 easily translates into:

Proposition 5.7. *Let $G = \operatorname{cl}\langle a_1, \ldots, a_d \rangle$ be a finitely generated powerful pro-p group. Writing $G_i := P_i(G)$ for $i \in \mathbb{N}$, the following assertions hold:*

(1) *G_i is powerfully embedded in G;*

(2) *$G_{i+k} = P_{k+1}(G_i) = G_i^{p^k}$ for each $k \in \mathbb{N}$, and in particular $G_{i+1} = \Phi(G_i)$;*

(3) *$G_i = G^{p^{i-1}} = \{x^{p^{i-1}} \mid x \in G\} = \operatorname{cl}\langle a_1^{p^{i-1}}, \ldots, a_d^{p^{i-1}} \rangle$;*

(4) *the map $x \mapsto x^{p^k}$ induces a homomorphism from G_i/G_{i+1} onto G_{i+k}/G_{i+k+1} for each $k \in \mathbb{N}$.*

Corollary 5.8. *If $G = \langle a_1, \ldots, a_d \rangle$ is a powerful pro-p group, then G decomposes as a product of its procyclic subgroups $\operatorname{cl}\langle a_i \rangle$, i.e. $G = \operatorname{cl}\langle a_1 \rangle \cdots \operatorname{cl}\langle a_d \rangle$.*

The *rank* of a profinite group G is defined to be the invariant $\operatorname{rk}(G) := \sup\{d(H) \mid H \leq_o G\}$. It can be shown that

$$\operatorname{rk}(G) = \sup\{d(H) \mid H \leq_c G\} = \sup\{\operatorname{rk}(G/N) \mid N \trianglelefteq_o G\};$$

see Exercise 6.3. Theorems 5.4 and 5.5 translate readily into:

Theorem 5.9 (Characterisation of pro-p groups of finite rank). *A pro-p group has finite rank if and only if it is finitely generated and virtually powerful.*

Moreover, if G is a finitely generated powerful pro-p group, then one has $\operatorname{rk}(G) = d(G)$.

The detailed proof of Theorem 5.5 also yields the following interesting 'local' description of pro-p groups of finite rank; cf. [10, Cor. 3.14].

Theorem 5.10. *Let G be a pro-p group and $r \in \mathbb{N}$. If every open subgroup of G contains an open normal subgroup $N \trianglelefteq_o G$ with $d(N) \leq r$, then G has finite rank.*

5.8   Pro-p groups of finite rank – summary of characterisations

There is a variety of other characterisations of the class of pro-p groups of finite rank. For instance, a pro-p group has finite rank if and only if it has polynomial subgroup growth; see Exercise 6.6. Considerably deeper, but most interesting, is the result that a pro-p group has finite rank if and only if it admits the structure of a p-adic Lie group. By way of a short summary, we record a small version of [10, Interlude A]:

Theorem 5.11 (Pro-p groups of finite rank – summary of characterisations). *Let G be a pro-p group. Then each of the following conditions is necessary and sufficient for G to have finite rank:*

(1) *G is finitely generated and virtually powerful;*

(2) *there exists $r \in \mathbb{N}$ such that every open subgroup of G contains an open normal subgroup $N \trianglelefteq_\mathrm{o} G$ with $d(N) \leq r$;*

(3) *G has polynomial subgroup growth;*

(4) *G is isomorphic to a closed subgroup of $\mathrm{GL}_d(\mathbb{Z}_p)$ for suitable $d \in \mathbb{N}$;*

(5) *G is a p-adic Lie group.*

We conclude this section by stating an intriguing problem which aims at yet another interesting characterisation of pro-p groups of finite rank. A profinite group G is said to be *noetherian* if it satisfies the ascending chain condition on closed subgroups.[4] It is easily seen that a pro-p group G is noetherian if and only if every closed subgroup of G is finitely generated. Consequently, every pro-p group of finite rank is noetherian. In fact, if G is a pro-p group of finite rank, then there is a uniform bound on the lengths of chains of closed subgroups $1 = G_0 \subseteq G_1 \subseteq \ldots \subseteq G_n = G$ with $|G_i : G_{i-1}| = \infty$ for all $i \in \{1, \ldots, n\}$. This bound is given by the dimension of G; see Section 7.1.

The following rather natural problem, which was posed by Lubotzky and Mann in [42, 43], has been open for more than 20 years.

Problem. *Does every noetherian pro-p group have finite rank?*

Suggestions for further reading

A detailed account of powerful p-groups is given in [10, Ch. 2], which includes full proofs of the stated results, also for $p = 2$. As mentioned in the introduction, the book [10] by Dixon, du Sautoy, Mann and Segal served as a close inspiration

---

[4]In point-set topology, it is customary to call a topological space noetherian if it satisfies the ascending chain condition on open subsets, but this notion is of little use in the context of profinite groups: a non-discrete profinite group never satisfies the ascending chain condition on open subsets. This should be contrasted with the observation that every profinite group satisfies trivially the ascending chain condition on open subgroups.

for large parts of these notes. Aspects of the theory of profinite groups, which are somewhat complementary to the topics covered in [10], can be found in the monograph [50], by Ribes and Zalesskii, and the book [60], by Wilson. In both books, one finds detailed accounts of how profinite groups arise as Galois groups, inverse limits, profinite completions and topological groups. Boston's chapter in [11] explains, from a group-theoretic point of view, the interactions between the theory of p-adic Galois representations and the structure of pro-p Galois groups; in particular, it contains a discussion of the important Fontaine–Mazur conjecture. Higgin's lecture notes [18] on topological groups put profinite groups in a broader perspective. An accessible introduction to the theorem of Nikolov and Segal mentioned in Section 5.5 is given in [53, Ch. 4], by Segal.

6   Second series of exercises

Exercise 6.1 (Procyclic groups and p-adic exponentiation).
(a) Recall that $\hat{\mathbb{Z}} = \varprojlim \mathbb{Z}/n\mathbb{Z}$ is the profinite completion of $\mathbb{Z}$. Show that the ring $\mathbb{Z}_p$ of p-adic integers is isomorphic to the pro-p completion of the ring $\mathbb{Z}$.
(b) Show that the profinite ring $\hat{\mathbb{Z}}$ decomposes as $\hat{\mathbb{Z}} = \prod_p \mathbb{Z}_p$.
(c) The profinite topology on $\mathbb{Z}$ is the topology whose open sets are the unions of cosets $a + b\mathbb{Z}$ with $a, b \in \mathbb{Z}$ and $b \neq 0$. Show that this agrees with the subspace topology coming from the inclusion $\mathbb{Z} \subseteq \hat{\mathbb{Z}}$. Note that every non-empty open subset of $\mathbb{Z}$ is infinite. Deduce from the equation $\{1, -1\} = \mathbb{Z} \setminus \bigcup \{p\mathbb{Z} \mid p \text{ prime}\}$ that there are infinitely many primes.
(d) Let G be a pro-p group. Let $g \in G$ and $\lambda = \sum_{k=0}^{\infty} a_k p^k \in \mathbb{Z}_p$. Write $\lambda_n := \sum_{k=0}^{n} a_k p^k$ to denote the partial sums, and show that the limit $\lim_{n \to \infty} g^{\lambda_n}$ exists. Denote this limit by g^λ, the λth *power* of g.
Let $g, h \in G$ and $\lambda, \mu \in \mathbb{Z}_p$. Convince yourself that p-adic exponentiation satisfies the common rules $g^{\lambda+\mu} = g^\lambda g^\mu$ and $g^{\lambda\mu} = (g^\lambda)^\mu$. Find a sufficient condition under which the equation $(gh)^\lambda = g^\lambda h^\lambda$ holds.
(e) Let G be a pro-p group and $g \in G$. Prove that p-adic exponentiation provides a surjective homomorphism $\mathbb{Z}_p \to \mathrm{cl}\langle g \rangle$, $\lambda \mapsto g^\lambda$.
Remark: A profinite group which is (topologically) generated by one element is called a *procyclic group*.
(f) Prove that a procyclic pro-p group is either finite and cyclic or isomorphic to $\mathbb{Z}_p$. Show more generally that a finitely generated abelian pro-p group G is isomorphic to $\mathbb{Z}_p^d \times F$ for some $d \in \mathbb{N}_0$ and a finite abelian p-group F. (*Hint:* Regard G as a finitely generated $\mathbb{Z}_p$-module.)

Exercise 6.2 (Explicit examples of pro-p groups).
(a) Let $d \in \mathbb{N}$. Prove that $\mathrm{GL}_d(\mathbb{Z}_p)$ is virtually a pro-p group. Can you guess a candidate for an open powerful pro-p subgroup of $\mathrm{GL}_d(\mathbb{Z}_p)$? (*Hint:* Consider the first congruence subgroup $\mathrm{GL}_d^1(\mathbb{Z}_p) = \{g \in \mathrm{GL}_d(\mathbb{Z}_p) \mid g \equiv 1 \pmod{p}\}$.)
(b) The Heisenberg group over $\mathbb{Z}_p$ is the group of upper uni-triangular 3×3 matrices over $\mathbb{Z}_p$. Work out the lower p-series of this group. Is it a powerful pro-p group? If not, is it of finite rank?

(c) Suppose that $p > 2$. Consider the Nottingham group G over $\mathbb{F}_p$, which was introduced in Exercise 4.3. Convince yourself that G is a topological group with respect to the subspace topology, inherited from $\mathbb{F}_p[\![t]\!]$. Show that G is a two-generated pro-p group. Is it powerful? If not, is it of finite rank? (*Hint:* Consider the abelianisations of its natural congruence subgroups $G_n := \{\mathbf{a} \in G \mid \mathbf{a} \equiv t \bmod t^{n+1}\}$.)

(d) Construct surjective homomorphisms from $C_p \wr C_{p^{n+1}}$ onto $C_p \wr C_{p^n}$ for all $n \in \mathbb{N}$. (*Hint:* Realise $C_p \wr C_{p^n}$ as the semidirect product of $\mathbb{F}_p[X]/(X^{p^n} - 1)$ by $\langle x \rangle \cong C_{p^n}$, with x acting as multiplication by X. Then convince yourself that there is a natural projection $\mathbb{F}_p[X]/(X^{p^{n+1}} - 1) \to \mathbb{F}_p[X]/(X^{p^n} - 1)$.) Set up a corresponding inverse system and take the inverse limit. The resulting group is the *pro-p wreath product* $C_p \hat{\wr} \mathbb{Z}_p$. Show that this group is two-generated but has infinite rank.

Remark: In a loose sense, $C_p \hat{\wr} \mathbb{Z}_p$ can be regarded as the smallest pro-p group which is not p-adic analytic; cf. [56].

Exercise 6.3 (Profinite groups: generating sets, universal property, rank).
(a) Let G be a profinite group and $X \subseteq G$. Show that the closure of X satisfies $\mathrm{cl}(X) = \bigcap\{XN \mid N \trianglelefteq_o G\}$. Deduce that X generates G if and only if X generates G modulo every open normal subgroup $N \trianglelefteq_o G$.

(b) Prove that every open subgroup of a finitely generated profinite group is finitely generated. (*Hint:* Use the corresponding result for abstract groups; namely, every finite-index subgroup of a finitely generated group is finitely generated.)

(c) Let $(G_i; \varphi_{ij})$ be an inverse system of groups based on a directed set I. Let $G = \varprojlim G_i$, and let $\varphi_i : G \to G_i$ denote the ith coordinate map. Show that $(G; \varphi_i)$ is characterised by the following universal property: given a group H and homomorphisms $\vartheta_i : H \to G_i$, $i \in I$, such that $\vartheta_i \varphi_{ij} = \vartheta_j$ whenever $i \succeq j$, there is a unique homomorphism $\vartheta : H \to G$ such that $\vartheta_i = \vartheta \varphi_i$ for all $i \in I$. (*Hint:* Start by drawing a corresponding diagram.)

(d) Convince yourself that every profinite group G is isomorphic to the inverse limit $\varprojlim G/N$ of its (continuous) finite quotients G/N, $N \trianglelefteq_o G$.

(e) Let G be a profinite group. Prove that $\mathrm{rk}(G) = \sup\{d(H) \mid H \leq_c G\} = \sup\{\mathrm{rk}(G/N) \mid N \trianglelefteq_o G\}$. For $N \trianglelefteq_c G$, show that $\max\{\mathrm{rk}(N), \mathrm{rk}(G/N)\} \leq \mathrm{rk}(G) \leq \mathrm{rk}(N) + \mathrm{rk}(G/N)$. Deduce that, if G has an open subgroup of finite rank, then G itself has finite rank.

Exercise 6.4 (Profinite groups: Sylow theory and finite images).
(a) Deduce from Tychonoff's theorem the following set-theoretical principle which frequently allows one to deduce properties of a profinite group from properties of its finite quotients: the inverse limit $\varprojlim X_i$ of an inverse system of non-empty finite sets X_i, $i \in I$, is non-empty. (*Hint:* Enforce the compatibility conditions in finite portions.)

(b) Let G be a profinite group. A *Sylow pro-p subgroup* of G is a maximal pro-p subgroup. Deduce from the Sylow theorems for finite groups that (i) every pro-p subgroup of G is contained in a Sylow pro-p subgroup and that (ii) any two Sylow pro-p subgroups of G are conjugate. (*Hint:* Use part (a) and take advantage of the fact that a profinite group is a pro-p group if and only if all its open subgroups have p-power index; see Exercise 6.5.)

(c) Prove that two finitely generated profinite groups are isomorphic if and only if they have the same class of finite groups as their finite (continuous) homomorphic images. (*Hint:* Set up a suitable inverse system of isomorphisms between finite (continuous) quotients of the two groups and use part (a).)

Give an example of two non-isomorphic pro-p groups which have the same class of finite groups as their finite (continuous) homomorphic images.

(d) Let Γ be a group and denote by $G := \varprojlim_{N \trianglelefteq_f \Gamma} \Gamma/N$ its profinite completion. Write $\vartheta : \Gamma \to G$ for the natural homomorphism $g \mapsto (gN)_N$. Show that ϑ induces an isomorphism $\Gamma/N \to G/\operatorname{cl}(N\vartheta)$ for each $N \trianglelefteq_f \Gamma$. Prove that every open subgroup of G is of the form $\operatorname{cl}(H\vartheta)$ where $H \leq_f \Gamma$.

Let Γ and Δ be finitely generated groups. Deduce from part (c) that their profinite completions $\hat{\Gamma}$ and $\hat{\Delta}$ are isomorphic if and only if Γ and Δ have the same class of finite groups as their finite homomorphic images.

Exercise 6.5 (Abstract finite-index subgroups of pro-p groups).

(a) Give an example of a pro-p group admitting finite-index subgroups which are not open. (*Hint:* Your group cannot be finitely generated.)

(b) Let G be a pro-p group and $H \leq_o G$. Show that $|G : H|$ is a power of p. (*Hint:* If $G = \varprojlim G_i$, think of G as subgroup of $\prod G_i$ and use that basic open subgroups of this latter group have p-power index.)

(c) Let G be a pro-p group and $H \leq_f G$. Show that $|G : H|$ is a power of p. (*Hint:* Replacing H by its core in G, you may assume that H is normal in G. Write $|G : H| = m = p^r q$ with $p \nmid q$, and put $X := \{g^m \mid g \in G\}$. Note that $X \subseteq H$ and that X is closed. Let $g \in G$. Show that $g^{p^r} \in XN$ for every $N \trianglelefteq_o G$. From $X = \bigcap \{XN \mid N \trianglelefteq_o G\}$ conclude that $g^{p^r} \in X$. Deduce that $|G : H| = p^r$.)

(d) Let G be a finitely generated pro-p group. Prove that the abstract commutator subgroup $[G, G]$ is closed, using the following fact about (abstract) nilpotent groups: if $\Gamma = \langle a_1, \dots, a_d \rangle$ is a nilpotent group, then every element of $[\Gamma, \Gamma]$ is equal to a product of the form $[x_1, a_1] \cdots [x_d, a_d]$ with $x_1, \dots, x_d \in \Gamma$. (*Hint:* Suppose that $G = \operatorname{cl}\langle a_1, \dots, a_d \rangle$ and consider $X := \{[g_1, a_1] \cdots [g_d, a_d] \mid g_1, \dots, g_d \in G\}$. Show that X is closed and that $X \equiv [G, G]$ modulo any open normal subgroup of G. Conclude that $[G, G] = X$ is closed.)

(e) Let G be a finitely generated pro-p group. According to part (d), the abstract commutator subgroup $[G, G]$ is closed. Observe that the abstract Frattini subgroup $G^p[G, G]$ can be written as $\{g^p \mid g \in G\}[G, G]$ and hence show that $G^p[G, G]$ is closed. Deduce that $G^p[G, G] = \Phi(G)$.

(f) Let G be a finitely generated pro-p group and $H \leq_f G$. Prove that H is open in G. (*Hint:* It is enough to prove the statement for normal subgroups. Arguing by induction on $|G : H|$, suppose that H is properly contained in G. Since $|G : H|$ is a p-power, $M := H\Phi(G) = HG^p[G, G]$ is a proper open subgroup of the group G. Note that M is finitely generated and apply induction to find that H is open in M.)

Exercise 6.6 (Pro-p groups with polynomial subgroup growth).
Let G be a finitely generated pro-p group, and for every $n \in \mathbb{N}_0$ let σ_n denote the number of open subgroups of index at most p^n in G. The group G is said to have *polynomial subgroup growth* (PSG) if there exist $c, \alpha \in \mathbb{R}$ such that $\sigma_n \leq cp^{n\alpha}$ for all $n \in \mathbb{N}_0$.

(a) Show that σ_n is finite for every $n \in \mathbb{N}$.

(b) Show that, if G has finite rank, then G has PSG. The remaining parts of the exercise are concerned with proving the converse.

(c) Let $r \in \mathbb{N}$, and suppose that $N \trianglelefteq_o G$ is maximal with the property $d(N) \geq r$. Show that N is equal to the centraliser of $N/\Phi(N)$ in G. (*Hint:* Write $C := C_G(N/\Phi(N)) \trianglelefteq_o G$ and assume for a contradiction that $N \subsetneq C$. Choose an element xN of order p in $C/N \cap \mathrm{Z}(G/N)$ and put $M := \langle x \rangle N \trianglelefteq_o G$. Deduce that $d(M) \geq d(N)$, in contradiction to $N \subsetneq M$.)

(d) Let V be a vector space of dimension d over $\mathbb{F}_p$. Show that every p-subgroup G of $\mathrm{GL}(V)$ admits a chain of normal subgroups

$$G = G_1 \supseteq G_2 \supseteq \ldots \supseteq G_{\lambda(d)} \supseteq G_{\lambda(d)+1} = 1$$

of length $\lambda(d) := \lceil \log_2(d) \rceil$ such that the quotients G_i/G_{i+1} of successive terms are elementary abelian. (*Hint:* It suffices to prove that a Sylow p-subgroup of $\mathrm{GL}(V)$ has a chain of normal subgroups of length $\lambda(d)$ such that the quotients of successive terms are elementary abelian. All Sylow p-subgroups of $\mathrm{GL}(V)$ are isomorphic to the group of upper uni-triangular matrices of degree d over $\mathbb{F}_p$.)

Show also that V contains at least $p^{(d-1)^2/4}$ subspaces of codimension $\lfloor d/2 \rfloor$.

(e) Let $r \in \mathbb{N}$, and let $N \trianglelefteq_o G$ be maximal with the property $d := d(N) \geq r$. Show that $|G : N| \leq p^{(r-1)\lambda(d)}$ where $\lambda(d) := \lceil \log_2(d) \rceil$. (*Hint:* By part (c), G/N acts faithfully by conjugation on $N/\Phi(N) \cong \mathbb{F}_p^d$. Note that every normal subgroup of G/N can be generated by $r - 1$ elements and use part (d).)

(f) Suppose that G has PSG and let $c, \alpha \in \mathbb{R}$ such that $\sigma_n \leq cp^{n\alpha}$ for all $n \in \mathbb{N}_0$. Show that there is a finite upper bound for the numbers $d(N)$ as N ranges over all open normal subgroups of G. (*Hint:* Let $r \in \mathbb{N}$, and suppose that $N \trianglelefteq_o G$ is maximal with the property $d := d(N) \geq r$. By considering suitable subgroups of $N/\Phi(N)$ derive from parts (d) and (e) that G contains at least $p^{(d-1)^2/4}$ open subgroups of index at most $p^{(r-1)\lambda(d)+\lfloor d/2 \rfloor}$. Use the fact that G has PSG to show that d, and hence r, is bounded above in terms of c and α.)

(g) Deduce from (f) and Theorem 5.10: if G has PSG, then G has finite rank.

Suggestions for further reading

For all exercises, one can consult [10]. As mentioned in Section 5.5, Exercise 6.5 is based on an exercise in Serre's classical text [54]. Exercise 6.6 touches upon the subject of subgroup growth, which is comprehensively treated in the monograph [44], by Lubotzky and Segal, and also discussed in Chapter III, by Voll.

7 Uniformly powerful pro-p groups and $\mathbb{Z}_p$-Lie lattices

7.1 Uniformly powerful pro-p groups

A finitely generated torsion-free powerful pro-p group is called a *uniformly powerful pro-p group*, or simply a *uniform pro-p group* for short. This concept and terminology is motivated by the following two results; see [10, Ch. 4].

Theorem 7.1 (Structure of finitely generated powerful pro-p groups). *Let G be a finitely generated powerful pro-p group. Then the elements of finite order in G form a characteristic subgroup T of G. Moreover, T is a powerful finite p-group and G/T is a uniform pro-p group. In particular, G is virtually uniform.*

Proposition 7.2 (Properties of uniform pro-p groups). *Let G be a finitely generated powerful pro-p group. Then the following are equivalent:*

(1) *G is uniform;*

(2) *for every $i \in \mathbb{N}$, the map $x \mapsto x^p$ induces an isomorphism from $P_i(G)/P_{i+1}(G)$ onto $P_{i+1}(G)/P_{i+2}(G)$;*

(3) *$d(H) = d(G)$ for every powerful open subgroup H of G.*

Let G be a pro-p group of finite rank. Then G contains an open uniform subgroup U. According to Proposition 7.2, the minimal number of generators for U does not depend on the particular choice of U and thus provides a useful invariant of G: the *dimension* of G is defined as $\dim(G) := d(U)$. One can show that for all $N \trianglelefteq_{\mathrm{c}} G$ one has

$$\dim(G) = \dim(N) + \dim(G/N).$$

The algebraically defined dimension of G is, in fact, the same as the dimension of G regarded as a p-adic Lie group; see [10, Theorem 8.36]. A first indication that G carries the structure of a p-adic analytic manifold is given by:

Proposition 7.3 (Multiplicative coordinate systems). *Let U be a uniform pro-p group and $d = d(U)$. Then every minimal generating set $\{a_1, \ldots, a_d\}$ for U yields a homeomorphism*

$$\mathbb{Z}_p^d \to U, \quad (\lambda_1, \ldots, \lambda_d) \mapsto a_1^{\lambda_1} \cdots a_d^{\lambda_d}.$$

This proposition can easily be proved from Proposition 7.2 and Corollary 5.8; see [10, Theorem 4.9]. The algebraic properties of the multiplicative coordinate systems are, however, not so good. We therefore set out to describe uniform pro-p groups in terms of more useful coordinate systems.

7.2   Associated additive structure

Let G be a uniform pro-p group of dimension d, and write $G_n := P_n(G) = G^{p^{n-1}}$ for the terms of the lower p-series of G. Our aim is to define on G the structure of an abelian group isomorphic to $\mathbb{Z}_p^d$. The new addition is to be defined canonically in terms of the original group multiplication and such that the two compositions agree on all abelian subgroups of G.

We take our inspiration from the formal identity

$$\exp(X + Y) = \lim_{n \to \infty} \left(\exp(X/n) \exp(Y/n) \right)^n,$$

known as the Lie product formula, which holds in the completed free associative algebra $\mathbb{Q}\langle\!\langle X, Y \rangle\!\rangle$ and can be traced back to the beginnings of Lie theory; cf. Exercise 9.1. Proposition 7.2 can be used to show that every element $x \in G_{n+1}$ admits a unique p^nth root in G, which we shall denote by $x^{p^{-n}}$. Moreover, the groups G_n admit larger and larger abelian quotients G_n/G_{2n} as $n \to \infty$. These crucial observations allow us to define the sum of $x, y \in G$ as

$$x + y := \lim_{n \to \infty} (x^{p^n} y^{p^n})^{p^{-n}}.$$

Essentially, we superimpose the groups G_n, by mapping them onto the reference set G, and notice that their composition maps become more and more alike as $n \to \infty$. Careful, but elementary considerations, cf. [10, §4.3], lead to:

Theorem 7.4 (Associated additive structure). *Let G be a uniform pro-p group of dimension d, and let $\{a_1, \ldots, a_d\}$ be a minimal generating set for G. Then the following hold:*

(1) *G with the operation $+$ constitutes a free $\mathbb{Z}_p$-module on the basis $\{a_1, \ldots, a_d\}$;*

(2) *the operation $+$ agrees with the original multiplication on all abelian subgroups of G;*

(3) *the terms of the lower p-series with respect to $+$ are the same as the ones for the original multiplication.*

In particular, this theorem implies:

- the neutral element of G with respect to $+$ equals the multiplicative identity element 1;

- inverses with respect to $+$ are the same as multiplicative inverses;

- p-adic exponentiation translates into scalar multiplication, i.e. $x^\lambda = \lambda x$ for all $x \in G$ and $\lambda \in \mathbb{Z}_p$.

A particularly useful consequence of the theorem is:

Corollary 7.5. *Let G be a uniform pro-p group of dimension d. Then the action of $\mathrm{Aut}(G)$ on G is $\mathbb{Z}_p$-linear with respect to the $\mathbb{Z}_p$-module structure on $(G, +)$. Moreover, $\mathrm{Aut}(G)$ embeds into $\mathrm{GL}_d(\mathbb{Z}_p)$ as a closed subgroup.*

The corollary implies in particular that the automorphism group of a pro-p group of finite rank is virtually again a pro-p group of finite rank. Another immediate consequence is that every pro-p group G of finite rank which contains an open uniform subgroup U with $\mathrm{Z}(U) = 1$ is linear over $\mathbb{Z}_p$. In fact, this is a special instance of Lazard's characterisation of p-adic analytic groups as linear groups over $\mathbb{Z}_p$, stated in Theorem 2.2.

7.3 Associated Lie structure

Let G be a uniform pro-p group, and write $G_n := P_n(G) = G^{p^{n-1}}$ for the terms of the lower p-series of G. Since all free $\mathbb{Z}_p$-modules of a given dimension are isomorphic, the procedure of passing from the uniform pro-p group G to the associated $\mathbb{Z}_p$-module $(G, +)$ inevitably involves a certain loss of information. More information can be saved by defining yet another operation, namely a Lie bracket. The new operation is to be defined canonically in terms of group commutators.

Again, we take our inspiration from a formal identity, namely

$$\exp(XY - YX) = \lim_{n \to \infty} \left(\exp(X/n)^{-1} \exp(Y/n)^{-1} \exp(X/n) \exp(Y/n) \right)^{n^2}$$

which holds in the completed free associative algebra $\mathbb{Q}\langle\!\langle X, Y \rangle\!\rangle$ and is intimately linked with Lie theory. Accordingly, we define the Lie bracket of $x, y \in G$ as

$$[x, y]_{\mathrm{Lie}} := \lim_{n \to \infty} [x^{p^n}, y^{p^n}]^{p^{-2n}}.$$

The individual terms make sense as $[G_n, G_n] \subseteq G_{2n}$, but, of course, one needs to check that the sequence converges. Careful, but elementary considerations, cf. [10, §4.5], lead to:

Theorem 7.6 (Associated Lie structure). *Let G be a uniform pro-p group. With the operation $[\cdot, \cdot]_{\mathrm{Lie}}$ the $\mathbb{Z}_p$-module $(G, +)$ becomes a $\mathbb{Z}_p$-Lie lattice.*

In the following, we denote the $\mathbb{Z}_p$-Lie lattice associated to G by $L(G)$. The next proposition assures us that the assignment of a Lie lattice to a uniform pro-p group is well behaved with respect to the passage to subgroups or quotients.

Proposition 7.7. *Let G be a uniform pro-p group. Let $H \leq_c G$ be a uniform subgroup, and let $N \trianglelefteq_c G$ such that G/N is uniform. Then N is uniform and:*

(1) *$L(H)$ constitutes a Lie sublattice of $L(G)$;*

(2) *$L(N)$ constitutes a Lie ideal of $L(G)$, the sets G/N and $L(G)/L(N)$ are equal and the natural epimorphism $G \to G/N$ of groups induces an epimorphism $L(G) \to L(G/N)$ of $\mathbb{Z}_p$-Lie lattices with kernel $L(N)$.*

Unlike the $\mathbb{Z}_p$-module $(G, +)$, the $\mathbb{Z}_p$-Lie lattice $L(G)$ actually captures all the information in the uniform pro-p group G. Indeed, our next task is to describe how the group multiplication can be recovered from the Lie bracket.

7.4   The Hausdorff formula

As mentioned in Section 2.4, the Hausdorff formula gives an expression for the formal power series

$$\Phi(X, Y) := \log(\exp(X) \cdot \exp(Y)) \in \mathbb{Q}\langle\!\langle X, Y \rangle\!\rangle$$

in non-commuting indeterminates X, Y; e.g. see [6, II.§6] or [10, §6.5]. In order to state the precise formula, we first note that the associative algebra $\mathbb{Q}\langle\!\langle X, Y \rangle\!\rangle$ admits in $[A, B] := AB - BA$ a natural Lie bracket. Expressing exp and log as power series, one can effectively eliminate the associative multiplication by a careful analysis and express $\Phi(X, Y)$ completely in terms of Lie commutators: $\Phi(X, Y) = \sum_{n=1}^{\infty} u_n(X, Y)$ is the infinite sum of homogeneous terms $u_n(X, Y)$ where

$$u_n(X, Y) = \sum_{m=1}^{n} \sum_{\substack{a_i, b_i \geq 0 \text{ s.t.} \\ a_i + b_i > 0, \\ \sum(a_i + b_i) = n}} \frac{(-1)^{m-1}}{mn \cdot a_1! b_1! \cdots a_m! b_m!} [_{a_1} X, _{b_1} Y, \ldots, _{a_m} X, _{b_m} Y]$$

with $[_{a_1} X, _{b_1} Y, \ldots, _{a_m} X, _{b_m} Y] = [\underbrace{X, \ldots, X}_{a_1}, \underbrace{Y, \ldots, Y}_{b_1}, \ldots, \underbrace{X, \ldots, X}_{a_m}, \underbrace{Y, \ldots, Y}_{b_m}]$ and all commutators being left-normed. A computation of the first three homogeneous terms $u_i(X, Y)$ shows that

$$\Phi(X, Y) = X + Y + \frac{[X, Y]}{2} + \frac{[X, Y, Y] - [X, Y, X]}{12} + \ldots$$

As it stands, the Hausdorff formula is an identity in formal power series. Let us explain its meaning as such. Consider the completed free associative algebra $A := \mathbb{Q}\langle\!\langle x_1, \ldots, x_d \rangle\!\rangle$ in d non-commuting indeterminates. Write $M := (x_1, \ldots, x_d)$ for the maximal ideal of A. It is easily seen that the exponential map and the logarithm map set up mutually inverse bijections between the sets M and $1 + M$. For $d = 1$, they even provide isomorphisms between the additive group M and the multiplicative group $1 + M$. But for $d \geq 2$ the groups M and $1 + M$ are clearly not isomorphic: M is abelian, whereas $1 + M$ is not. The situation can be saved by equipping M with the commutator Lie bracket: the Hausdorff formula shows that the multiplicative group $1 + M$ can be described entirely in terms of the Lie algebra M.

We want to use the Hausdorff formula to recover a uniform pro-p group G from the associated $\mathbb{Z}_p$-Lie lattice $L(G)$. Naturally, this situation is more complicated. For instance, the question of convergence has to be considered more seriously.

7.5 Applying the Hausdorff formula

A $\mathbb{Z}_p$-Lie lattice L is *powerful* if p is odd and $[L, L] \subseteq pL$, or if $p = 2$ and $[L, L] \subseteq 4L$. It is easily seen that the Lie lattice $L(G)$ associated to a uniform pro-p group is powerful. It is also worth noting that for any $\mathbb{Z}_p$-Lie lattice L the sublattice pL (respectively $4L$) is powerful if p is odd (respectively $p = 2$).

Let L be a powerful $\mathbb{Z}_p$-Lie lattice and let $x, y \in L$. A suitable analysis of the p-adic valuations of the rational coefficients which appear in the homogeneous components $u_n(X, Y)$ of the Hausdorff formula shows that $u_n(x, y) \in L$ for all $n \in \mathbb{N}$. Moreover, the sequence $u_n(x, y) \to 0$ as $n \to \infty$. Consequently, the limit $\Phi(x, y) := \sum_{n=1}^{\infty} u_n(x, y)$ exists in L. The formal properties of the logarithm and exponential series imply:

Theorem 7.8. *If L is a powerful $\mathbb{Z}_p$-Lie lattice, then the Hausdorff formula induces a group structure on L, with multiplication given by $xy = \Phi(x, y)$. The resulting group is a uniform pro-p group.*

Exercise 9.1 illustrates that some care is required in applying formal identities. We refer to [10, §9.4] for details. One can check that, if the construction is applied to the Lie lattice $L(G)$ associated to a uniform pro-p group G, one recovers the original group G. The assignment $G \mapsto L(G)$ thus defines an equivalence between the category of uniform pro-p groups and the category of powerful $\mathbb{Z}_p$-Lie lattices.

This equivalence in turn induces a functor from the category of pro-p groups of finite rank (which is equal to the category of p-adic analytic pro-p groups) to the category of finite-dimensional $\mathbb{Q}_p$-Lie algebras, taking G to $\mathcal{L}(G) := \mathbb{Q}_p \otimes_{\mathbb{Z}_p} L(U)$ where U is any open uniform subgroup of G. Likewise, the image under the functor of a homomorphism between two pro-p groups of finite rank only depends on its restriction to an open uniform subgroup.

Suggestions for further reading

The key reference for uniformly powerful pro-p groups and their relation to powerful $\mathbb{Z}_p$-Lie lattices is the book [10], by Dixon, du Sautoy, Mann and Segal. The theory of Lie groups over non-archimedean local fields was originally developed by Bourbaki, Lazard and Serre; e.g. see [6, 55]. Lazard's seminal paper [39], which describes how the group-theoretic properties of a pro-p group reflect its status as a p-adic analytic group, continues to be a valuable inspiration. Reutenauer's book [49] on free Lie algebras may be of interest, for instance, in connection with the the Hausdorff formula. There are many fine books on real Lie groups, including modern introductions such as [36], by Kosmann-Schwarzbach, and [52], by Rossmann. The classical text [9], by Chevalley, is also highly recommended.

8 The group $\mathrm{GL}_d(\mathbb{Z}_p)$, just-infinite pro-$p$ groups and the Lie correspondence for saturable pro-p groups

8.1 The group $\mathrm{GL}_d(\mathbb{Z}_p)$ – an example

Let $d \in \mathbb{N}$. In order to illustrate the abstract concepts introduced in Section 7, we discuss in some detail the group $\mathrm{GL}_d(\mathbb{Z}_p)$. Clearly, $\mathrm{GL}_d(\mathbb{Z}_p)$ can be regarded as a topological group with respect to the p-adic topology, i.e. with respect to the subspace topology induced from the natural topology on the space $\mathrm{Mat}_d(\mathbb{Z}_p)$ of all $d \times d$ matrices over $\mathbb{Z}_p$. The principal congruence subgroups

$$G_i := \mathrm{GL}_d^i(\mathbb{Z}_p) := \{g \in \mathrm{GL}_d(\mathbb{Z}_p) \mid g \equiv 1 \pmod{p^i}\}, \quad i \in \mathbb{N}_0,$$

provide a natural filtration of $\mathrm{GL}_d(\mathbb{Z}_p)$. For each $i \in \mathbb{N}_0$, the ith principal congruence subgroup G_i is equal to the kernel of the natural projection $\mathrm{GL}_d(\mathbb{Z}_p) \to \mathrm{GL}_d(\mathbb{Z}/p^i\mathbb{Z})$ and hence forms an open normal subgroup of $\mathrm{GL}_d(\mathbb{Z}_p)$. Note that a matrix $x \in \mathrm{Mat}_d(\mathbb{Z}_p)$ is invertible if and only if it is invertible modulo p. This yields yet another description of the principal congruence subgroups: one has $G_0 = \mathrm{GL}_d(\mathbb{Z}_p)$ and $G_i = 1 + p^i \mathrm{Mat}_d(\mathbb{Z}_p)$ for each $i \geq 1$. In particular, it follows that

$$|G_0 : G_1| = |\mathrm{GL}_d(\mathbb{F}_p)| = (p^d - 1)(p^d - p) \cdots (p^d - p^{d-1}),$$

$$|G_1 : G_i| = p^{d^2(i-1)} \quad \text{for } i \geq 1.$$

Moreover, the groups G_i form a base of open neighbourhoods for the identity matrix in $\mathrm{Mat}_d(\mathbb{Z}_p)$ and thus determine completely the topology on $\mathrm{GL}_d(\mathbb{Z}_p)$: every open neighbourhood of 1 in $\mathrm{GL}_d(\mathbb{Z}_p)$ contains one of the open normal subgroups G_i. It follows that $\mathrm{GL}_d(\mathbb{Z}_p)$ is profinite and that G_1 is a pro-p group. Put $\varepsilon := 0$ if p is odd, $\varepsilon := 1$ if $p = 2$; and set $G := G_{1+\varepsilon}$.

Proposition 8.1. *The group $G = \mathrm{GL}_d^{1+\varepsilon}(\mathbb{Z}_p)$ is a uniform pro-p group and $\dim(G) = \mathrm{rk}(G) = d(G) = d^2$. Moreover, the lower p-series of G coincides with the natural congruence filtration, i.e. $P_i(G) = G_{i+\varepsilon} = \mathrm{GL}_d^{i+\varepsilon}(\mathbb{Z}_p)$ for all $i \in \mathbb{N}$.*

Sketch of proof for $p > 2$. As p is odd, we have $G = G_1$. An easy computation shows that every quotient G_i/G_{i+1} of successive terms of the congruence filtration G_i, $i \in \mathbb{N}$, constitutes an elementary p-group of rank d^2, which is central in G/G_{i+1}. Thus $P_i(G) \subseteq G_i$ for all $i \in \mathbb{N}$. Below we show that $G_2 = \{x^p \mid x \in G\}$. This implies that $P_2(G) = G_2 = G^p$, hence G is powerful. Next we conclude from Proposition 5.7 that $P_i(G)/P_{i+1}(G)$ is an elementary p-group of rank at most d^2 for every $i \geq 2$. In view of the inclusions $P_i(G) \subseteq G_i$, it follows that $P_i(G) = G_i$ for all $i \in \mathbb{N}$ and that G is uniform of dimension $\dim(G) = d^2$, as wanted.

It remains to prove that every element of G_2 is a pth power of an element of the group G. In other words, given $A \in \mathrm{Mat}_d(\mathbb{Z}_p)$ we are to solve

$$(1 + pX)^p = 1 + p^2 A \quad \text{with } X \in \mathrm{Mat}_d(\mathbb{Z}_p).$$

We construct a solution X by means of successive approximations $X_i \in \mathrm{Mat}_d(\mathbb{Z}_p)$ modulo p^i, $i \in \mathbb{N}$, similarly as in Exercise 4.4. These approximations X_i will form a convergent sequence whose limit X will be an exact solution. Some care has to be taken, because matrix multiplication in general is not commutative. But we will construct each X_i so that it commutes with the given matrix A.

Set $X_1 := X_2 := X_3 := A$ and note that $(1 + pA)^p \equiv 1 + p^2 A$ modulo p^3. Now let $i \geq 4$ and suppose, inductively, that we have found a matrix X_{i-1}, commuting with A, such that $(1 + pX_{i-1})^p \equiv 1 + p^2 A$ modulo p^{i-1}. Then

$$(1 + pX_{i-1})^p = 1 + p^2 A + p^{i-1} E \qquad \text{for some } E \in \mathrm{Mat}_d(\mathbb{Z}_p).$$

Observe that E commutes with A and X_{i-1}. Put $X_i := X_{i-1} - p^{i-3} E$. Then X_i commutes with A, and a short computation shows that, modulo p^i, one has

$$\begin{aligned}
(1 + pX_i)^p &= (1 + pX_{i-1} - p^{i-2} E)^p \\
&\equiv (1 + pX_{i-1})^p - p(1 + pX_{i-1})^{p-1} p^{i-2} E \\
&\equiv 1 + p^2 A + p^{i-1} E - p^{i-1} E \\
&\equiv 1 + p^2 A.
\end{aligned}$$

$\square$

According to Section 7, there is a natural $\mathbb{Z}_p$-Lie lattice $L(G)$ associated to the uniform pro-p group G. Consider the $\mathbb{Z}_p$-Lie lattice $\mathfrak{gl}_d(\mathbb{Z}_p)$ of all $d \times d$ matrices over $\mathbb{Z}_p$, subject to the commutator Lie bracket. Similarly as the group $\mathrm{GL}_d(\mathbb{Z}_p)$, the Lie lattice $\mathfrak{gl}_d(\mathbb{Z}_p)$ admits a natural congruence filtration

$$\mathfrak{gl}_d^i(\mathbb{Z}_p) := \{x \in \mathfrak{gl}_d(\mathbb{Z}_p) \mid x \equiv 0 \pmod{p^i}\} = p^i \mathfrak{gl}_d(\mathbb{Z}_p), \quad i \in \mathbb{N}_0.$$

Put $\mathfrak{g} := \mathfrak{gl}_d^{1+\varepsilon}(\mathbb{Z}_p)$. Clearly, $\mathfrak{g}$ is a powerful $\mathbb{Z}_p$-Lie lattice.

Proposition 8.2. *The $\mathbb{Z}_p$-Lie lattice $L(G)$ associated to the uniform pro-p group $G = \mathrm{GL}_d^{1+\varepsilon}(\mathbb{Z}_p)$ is isomorphic to $\mathfrak{g} = \mathfrak{gl}_d^{1+\varepsilon}(\mathbb{Z}_p)$.*

Sketch of proof for $p > 2$. The correspondence between $G = \mathrm{GL}_d^1(\mathbb{Z}_p)$ and $\mathfrak{g} = \mathfrak{gl}_d^1(\mathbb{Z}_p)$ admits an explicit interpretation through the logarithm and the exponential map. For instance, one can check directly that the Lie bracket obtained in the construction of $L(G)$ is the same as the one of $\mathfrak{g}$, if one passes from one Lie lattice to the other by means of the logarithm and the exponential map.

Concretely, one may proceed as follows. A natural $\mathbb{Z}_p$-basis for $\mathfrak{g}$ is given by the p-multiples of the d^2 elementary matrices, i.e. by the matrices with one entry equal to p and all remaining entries equal to 0. One can explicitly compute the images of these basis elements in G under the exponential map. For any two basis elements $a, b \in \mathfrak{g}$, one can then verify that $\exp(ab - ba)$ is the same as the value which results from the corresponding limit formula, with input $x := \exp(a)$ and $y := \exp(b)$, provided in Section 7.3. $\square$

8.2 Just-infinite pro-p groups

A profinite group is *just-infinite* if it is infinite but admits no proper infinite quotients. It is easily seen that every just-infinite pro-p group is finitely generated and that every infinite finitely generated pro-p group has a just-infinite homomorphic image; see Exercise 9.4. Just-infinite pro-p groups play a similar role in the theory of pro-p groups as finite simple groups in the theory of finite groups. Many of the better-known just-infinite pro-p groups are groups of Lie type, defined over $\mathbb{Z}_p$ or over the pro-p ring $\mathbb{F}_p[\![t]\!]$ of formal power series with coefficients in $\mathbb{F}_p$. In addition, there are several interesting 'exceptional' examples of just-infinite pro-p groups, such as the Nottingham group; see Exercise 9.4. As yet no convincing proposal has been put forward for classifying just-infinite pro-p groups. In fact, one can construct uncountably many pairwise non-isomorphic just-infinite pro-p groups; see [32, Ch. XIII]. So a first step would be to give a precise and sensible meaning to the word 'classification' in the given context.

Best understood among the just-infinite pro-p groups are the p-adic analytic ones. Every soluble just-infinite pro-p group is virtually abelian and hence p-adic analytic. Indeed, the soluble just-infinite pro-p groups are irreducible p-adic space groups, and they can be investigated by the methods developed to study pro-p groups of finite coclass. The non-soluble p-adic analytic just-infinite pro-p groups can be realised as open subgroups of the groups $\mathbf{G}_{\mathbb{Q}_p}$ of $\mathbb{Q}_p$-rational points of certain semisimple algebraic groups defined over the field $\mathbb{Q}_p$. It is this description which makes them accessible in a rather explicit way; see [32].

Indeed, the non-soluble p-adic analytic just-infinite pro-p groups can naturally be partitioned into commensurability classes, where two profinite groups are *commensurable* if they have isomorphic open subgroups. One can then show that within each commensurability class of non-soluble p-adic analytic just-infinite pro-p groups there is (up to isomorphism) a unique *maximal* representative G which has the property that every just-infinite pro-p group which is commensurable to G embeds as an open subgroup into G.

The maximal group G which is commensurable to a given non-soluble p-adic analytic just-infinite pro-p group H can be constructed as follows. To H one associates via an open uniform subgroup $U \leq_o H$ the $\mathbb{Q}_p$-Lie algebra $\mathcal{L}(H) = \mathbb{Q}_p \otimes_{\mathbb{Z}_p} L(U)$. This Lie algebra turns out to be the direct sum of p^e copies of a simple $\mathbb{Q}_p$-Lie algebra for a suitable $e \in \mathbb{N}_0$, with $e = 0$ corresponding to the most interesting case. The automorphism group of the Lie algebra $\mathcal{L}(H)$ can be regarded as an algebraic group $\mathbf{G}$ defined over $\mathbb{Q}_p$. We remark that the classification of simple $\mathbb{Q}_p$-Lie algebras and simple algebraic groups over $\mathbb{Q}_p$ can be used to obtain an overview of the groups that occur. Since H is non-soluble and just-infinite, it acts faithfully on $\mathcal{L}(H)$ and thus embeds into the group $\mathbf{G}_{\mathbb{Q}_p}$ of $\mathbb{Q}_p$-rational points. Being a pro-p group, H is contained in a Sylow pro-p subgroup G of $\mathbf{G}_{\mathbb{Q}_p}$. A suitable Sylow theorem implies that all Sylow pro-p subgroups of $\mathbf{G}_{\mathbb{Q}_p}$ are conjugate. From this one shows that G is a maximal just-infinite pro-p group within the commensurability class of H.

It can be shown that a non-soluble p-adic analytic just-infinite pro-p group is never isomorphic to a proper subgroup of itself; cf. [32, Ch. III]. In contrast to this, the known just-infinite pro-p groups which are not p-adic analytic do admit proper subgroups which are isomorphic to the original groups. This leads to the interesting:

Problem. *Suppose that G is a just-infinite pro-p group which is not isomorphic to any of its proper closed subgroups $H <_c G$. Does it follow that G is p-adic analytic?*

8.3   Potent filtrations and saturable pro-p groups

In his seminal paper [39], entitled *Groupes analytiques p-adiques*, Lazard develops the theory of p-adic Lie groups from a class of groups which he calls 'groupes p-saturables'. These saturable pro-p groups include uniform pro-p groups, but form a strictly larger class; see [35]. From a group-theoretic perspective saturable pro-p groups are, however, not as comfortable to work with as uniform pro-p groups. Intuitively, a pro-p group G is saturable if we can associate to it a $\mathbb{Z}_p$-Lie lattice $L(G)$ via the limit process described in Section 7. González-Sánchez has developed a useful description of saturable pro-p groups in terms of potent filtrations; see [13].

Let G be a pro-p group, and let N be a closed normal subgroup of G. A *potent filtration* of N in G is a descending series N_i, $i \in \mathbb{N}$, of closed normal subgroups of G such that (i) $N_1 = N$, (ii) $\bigcap\{N_i \mid i \in \mathbb{N}\} = 1$, (iii) $[N_i, G] \subseteq N_{i+1}$ and $[N_{i, p-1} G] \subseteq N_{i+1}^p$ for all $i \in \mathbb{N}$. We say that N is *PF-embedded* in G if there exists a potent filtration of N in G. The group G is a *PF-group* if G is PF-embedded in itself.

To ease notation, group-theoretic constructs within topological groups will from now on be implicitly geared toward closed subgroups. For instance, if H, K are closed subgroups of a topological group G, we interpret $[H, K]$ as the *closed* subgroup generated by all commutators $[h, k]$ with $h \in H$ and $k \in K$.

Some basic properties of PF-embedded subgroups, which are listed in the next lemma, follow essentially from the Hall–Petrescu collection formula. This formula states that for elements x, y of any group G and $n \in \mathbb{N}$ one has

$$x^n y^n = (xy)^n c_2^{\binom{n}{2}} \cdots c_i^{\binom{n}{i}} \cdots c_{n-1}^n c_n \qquad \text{for suitable } c_i \in \gamma_i(G),\ i \in \{2, \dots, n\};$$

see [21, III.§9.4].

Lemma 8.3 (Properties of PF-embedded subgroups). *Let G be a pro-p group, and let N, M be PF-embedded subgroups of G. Then:*

(1) *NM, N^p and $[N, _k G]$ are PF-embedded in G for all $k \in \mathbb{N}$;*

(2) *$[N^p, G] = [N, G]^p$;*

(3) *$N^p = \{x^p \mid x \in N\}$;*

(4) *if G is torsion-free and $x^p \in N^p$, then $x \in N$; moreover, if $x, y \in N$ such that $x^p = y^p$, then $x = y$.*

González-Sánchez' characterisation of saturable pro-p groups in [13] is:

Theorem 8.4 (Saturable pro-p groups as PF-groups)**.** *Let G be a torsion-free finitely generated pro-p group. Then G is saturable if and only if G – or equivalently $G/\Phi(G)^p$ – is a PF-group.*
In particular, if $\gamma_p(G) \subseteq \Phi(G)^p$, then G is saturable.

It is not difficult to check that every uniform pro-p group G satisfies $\gamma_p(G) \subseteq \Phi(G)^p$. Hence uniform pro-$p$ groups are saturable. In fact, if G is a torsion-free finitely generated pro-p group satisfying $\gamma_p(G) \subseteq \Phi(G)^p$, then the lower p-series of G provides a potent filtration; see Exercise 9.5.

8.4   Lie correspondence

One difficulty in working with uniform pro-p groups is that the property of being powerful is not coherently inherited by subgroups; see Exercise 9.2. For instance, this causes problems if one tries to set up a Lie correspondence between subgroups of a uniform pro-p group and Lie sublattices of the corresponding $\mathbb{Z}_p$-Lie lattice. The situation improves substantially if instead one works with saturable pro-p groups. Using Theorem 8.4, González-Sánchez and Klopsch proved, in [15]:

Theorem 8.5. *Every torsion-free p-adic analytic pro-p group of dimension less than p is saturable. On the other hand, there exists a torsion-free p-adic analytic pro-p group of dimension p which is not saturable.*

This allows one to study torsion-free p-adic analytic pro-p groups of dimension less than p by means of $\mathbb{Z}_p$-Lie lattices, similarly as finite p-groups of nilpotency class less than p can be investigated based on the Lazard correspondence. The technique is particularly useful if one starts from a globally defined arithmetic group, such as $\mathrm{SL}_d(\mathbb{Z})$, and considers pro-$p$ subgroups of its completions at primes p, such as Sylow pro-p subgroups of $\mathrm{SL}_d(\mathbb{Z}_p)$. For fixed d, for almost all p, these local groups are torsion-free and saturable. In addition, González-Sánchez and Klopsch proved:

Proposition 8.6. *Let G be a saturable pro-p group, and let $H \leq_c G$ such that $\dim(H) \leq p$. Then H is saturable and hence corresponds to a Lie sublattice $L(H)$ of $L(G)$.*

This proposition gives a conceptually satisfying approach to setting up a Lie correspondence for subgroups of a saturable pro-p group; see [15]. A similar correspondence in the context of uniform pro-p groups was originally discovered and proved in [23], by Ilani, via *ad-hoc* arguments.

Theorem 8.7 (Lie correspondence)**.** *Let G be a saturable pro-p group and let $L(G)$ be the associated saturable $\mathbb{Z}_p$-Lie lattice. Suppose that $K, H \subseteq_c G$ are closed subsets of G, and denote them by $L(K)$, $L(H)$ when regarded as subsets of $L(G)$.*

(1) *Suppose that H is a subgroup of G and that $\dim\langle x, y\rangle_{\mathrm{Grp}} \leq p$ for all $x, y \in H$. Then $L(H)$ is a Lie sublattice of $L(G)$. Moreover, if K is a normal subgroup of H, then $L(K)$ is a Lie ideal of $L(H)$.*

(2) *Suppose that $L(H)$ is a Lie sublattice of $L(G)$ and that $\dim\langle x, y\rangle_{\mathrm{Lie}} \leq p$ for all $x, y \in L(H)$. Then H is a subgroup of G. Moreover, if $L(K)$ is a Lie ideal of $L(H)$, then K is a normal subgroup of H.*

Theorem 8.7 has natural applications, for instance to the subject of subgroup growth; cf. [44]. Indeed, it forms the basis for studying the subgroup zeta functions of p-adic analytic pro-p groups, such as $\mathrm{GL}_d^1(\mathbb{Z}_p)$, via their associated Lie lattices. It remains a challenging problem to describe the subgroup growth of the analytic pro-p groups $\mathrm{GL}_d^1(\mathbb{Z}_p)$, $d \in \mathbb{N}$. At least for $p \geq d^2$ this problem 'reduces' to understanding the sublattice growth of the $\mathbb{Z}_p$-Lie lattice $\mathfrak{gl}_d^1(\mathbb{Z}_p)$.

Suggestions for further reading

The group $\mathrm{GL}_d(\mathbb{Z}_p)$ is also treated as an example in [10, Ch. 5], by Dixon, du Sautoy, Mann and Segal. For more on just-infinite pro-p groups, we refer to [32], by Klaas, Leedham-Green and Plesken, and Wilson's chapter in [11] on just-infinite abstract and profinite groups. The former contains a detailed study of p-adic analytic just-infinite pro-p groups. As for saturable pro-p groups and the Lie correspondence, we refer to Lazard's paper [39] and the papers [35, 13, 15] already referred to in the main text. Additional, 'functorial' accounts of p-adic Lie theory are given in [10, Ch. 9], as well as in [6, III.§3-4], by Bourbaki, and [55, Part II], by Serre. It is interesting to compare the results in the p-adic setting with the Lie theory for real Lie groups; e.g. see [52, Ch. 2]. For subgroup zeta functions of groups, we refer to Chapter III, by Voll, and the references given therein.

9 Third series of exercises

Exercise 9.1 (Exponential and logarithm series).
(a) Prove the Lie product formula

$$\exp(X + Y) = \lim_{n \to \infty} \left(\exp(X/n) \exp(Y/n) \right)^n,$$

which holds in the completed free associative algebra $\mathbb{Q}\langle\langle X, Y\rangle\rangle$. (*Hint:* Start heuristically with $\exp(\varepsilon X) \exp(\varepsilon Y) = \exp(\varepsilon(X + Y)) + o(\varepsilon)$.)
(b) Show that in the 2-adic numbers $\mathbb{Q}_2$ one has $\log(-1) = 0 = \log(1)$. Conclude that $\exp(\log(-1)) = 1$ is not equal to -1!

Exercise 9.2 (The special linear groups $\mathrm{SL}_d(\mathbb{Z}_p)$).
(a) Let $d \in \mathbb{N}$ and consider the topological group $\mathrm{SL}_d(\mathbb{Z}_p)$. Show that this group is virtually a pro-p group and display an open uniform subgroup, together with its lower p-series. Realise the associated powerful $\mathbb{Z}_p$-Lie lattice explicitly as a Lie sublattice of $\mathfrak{gl}_d(\mathbb{Z}_p)$.

(b) Show that every open neighbourhood of 1 in $\mathrm{SL}_2(\mathbb{Z}_p)$ contains an open subgroup which is not powerful.

Exercise 9.3 (The quaternion group $\mathrm{SL}_1(\Delta_p)$).
Suppose that $p > 2$, and let $\rho \in \{1, 2, \ldots, p-1\}$ be a non-square modulo p. The 4-dimensional *quaternion algebra* over $\mathbb{Q}_p$ is the associative algebra

$$\mathbb{D}_p := \mathbb{Q}_p + \mathbb{Q}_p \mathbf{u} + \mathbb{Q}_p \mathbf{v} + \mathbb{Q}_p \mathbf{uv},$$

defined by the multiplication rules

$$\mathbf{u}^2 = \rho, \quad \mathbf{v}^2 = p, \quad \mathbf{uv} = -\mathbf{vu}.$$

The *reduced norm* and the *reduced trace* of an element $\mathbf{x} = \alpha + \beta \mathbf{u} + \gamma \mathbf{v} + \delta \mathbf{uv} \in \mathbb{D}_p$ are given by

$$\mathrm{N}(\mathbf{x}) = \alpha^2 - \rho\beta^2 - p\gamma^2 + \rho p \delta^2 \quad \text{and} \quad \mathrm{T}(\mathbf{x}) = 2\alpha.$$

We write $\mathrm{SL}_1(\mathbb{D}_p) := \{\mathbf{x} \in \mathbb{D}_p \mid \mathrm{N}(\mathbf{x}) = 1\}$ and $\mathfrak{sl}_1(\mathbb{D}_p) := \{\mathbf{x} \in \mathbb{D}_p \mid \mathrm{T}(\mathbf{x}) = 0\}$.

(a) Show that $\mathbb{D}_p$ is a skew field. (*Hint:* Use the norm map.)
(b) Prove that $\mathrm{SL}_1(\mathbb{D}_p)$ is a compact topological group. (*Hint:* Consider $\mathbf{x} = \alpha + \beta \mathbf{u} + \gamma \mathbf{v} + \delta \mathbf{uv} \in \mathbb{D}_p$ with $\mathrm{N}(\mathbf{x}) = 1$. Note that $v_p(\alpha^2 - \rho\beta^2)$ is even, while $v_p(p\gamma^2 - \rho p \delta^2)$ is odd. Conclude that $\alpha, \beta, \gamma, \delta \in \mathbb{Z}_p$. Now use the fact that the norm map is continuous.)
Remark: The group $\mathrm{SL}_2(\mathbb{Q}_p)$, in contrast, is clearly not compact.
(c) Show that $\mathfrak{sl}_1(\mathbb{D}_p)$ is a 3-dimensional simple $\mathbb{Q}_p$-Lie algebra. Prove that $\mathfrak{sl}_1(\mathbb{D}_p)$ does not have any subalgebras of dimension 2. Conclude that $\mathfrak{sl}_1(\mathbb{D}_p)$ is not isomorphic to the Lie algebra $\mathfrak{sl}_2(\mathbb{Q}_p)$.
Remark: There are (up to isomorphism) precisely two 3-dimensional simple $\mathbb{Q}_p$-Lie algebras, namely $\mathfrak{sl}_1(\mathbb{D}_p)$ and $\mathfrak{sl}_2(\mathbb{Q}_p)$.
(d) Note that $\Delta_p := \mathbb{Z}_p + \mathbb{Z}_p \mathbf{u} + \mathbb{Z}_p \mathbf{v} + \mathbb{Z}_p \mathbf{uv}$ constitutes a $\mathbb{Z}_p$-order of $\mathbb{D}_p$, i.e. a $\mathbb{Z}_p$-subalgebra whose $\mathbb{Q}_p$-span is equal to the entire algebra $\mathbb{D}_p$. Show that Δ_p admits a unique maximal ideal $\mathfrak{p}$, which is generated by $\mathbf{v}$.
Remark: One can extend the p-adic valuation on $\mathbb{Q}_p$ uniquely to a valuation on the skew field $\mathbb{D}_p$. The element $\mathbf{v}$ is a uniformiser for this valuation, i.e. it plays a similar role as p does for the valuation on $\mathbb{Q}_p$.
(e) Write $\mathfrak{sl}_1(\Delta_p) := \mathfrak{sl}_1(\mathbb{D}_p) \cap \Delta_p$ and $\mathbf{i} := \frac{1}{2}\mathbf{u}$, $\mathbf{j} := \frac{1}{2}\mathbf{v}$, $\mathbf{k} := \frac{1}{2}\mathbf{uv}$. Note that $\mathfrak{sl}_1(\Delta_p) = \mathbb{Z}_p \mathbf{i} + \mathbb{Z}_p \mathbf{j} + \mathbb{Z}_p \mathbf{k}$ and work out the commutators $[\mathbf{i}, \mathbf{j}]$, $[\mathbf{i}, \mathbf{k}]$, $[\mathbf{j}, \mathbf{k}]$ in terms of the new basis $\mathbf{i}, \mathbf{j}, \mathbf{k}$.
(f) Note that $\mathfrak{sl}_1^2(\Delta_p) := p \, \mathfrak{sl}_1(\Delta_p)$ is powerful. Convince yourself that the corresponding uniform pro-p group, which is defined via the Hausdorff formula, is equal to the group $\mathrm{SL}_1^2(\Delta_p) := \mathrm{SL}_1(\mathbb{D}_p) \cap (1 + p\Delta_p)$.
Conclude that $\mathrm{SL}_1(\mathbb{D}_p)$ is a 3-dimensional just-infinite compact p-adic analytic group which is not commensurable with $\mathrm{SL}_2(\mathbb{Z}_p)$.

Exercise 9.4 (Just-infinite pro-p groups).
(a) Prove that every just-infinite pro-p group is finitely generated.

Remark: I do not know whether there are just-infinite profinite groups which are not finitely generated.

(b) Let G be a pro-p group of finite rank with open uniform subgroup $U \leq_o G$. Prove that, if the associated $\mathbb{Q}_p$-Lie algebra $\mathcal{L}(G) = \mathbb{Q}_p \otimes_{\mathbb{Z}_p} L(U)$ is simple, then G is just-infinite.

(c) Determine all abelian just-infinite pro-p groups. Show that every soluble just-infinite pro-p group is virtually abelian. Construct a soluble just-infinite pro-p group which is not abelian.

(d) Give an example of an infinite pro-p group which does not have any just-infinite quotients. (*Hint:* Your group cannot be finitely generated.) In contrast, show that every infinite finitely generated pro-p group G admits a just-infinite quotient. (*Hint:* Consider an ascending chain $N_1 \subseteq N_2 \subseteq \ldots \subseteq G$ of closed normal subgroups of G such that G/N_i is infinite for all $i \in \mathbb{N}$, and assume for a contradiction that $M := \mathrm{cl}(\bigcup\{N_i \mid i \in \mathbb{N}\})$ is open in G. Conclude that M is finitely generated and derive a contradiction. Now apply Zorn's lemma.)

(e) Consider the profinite group $G := \prod_p C_p$, where the product extends over all primes p. Prove that G is finitely generated, but does not admit any just-infinite quotient.

(f) Suppose that $p > 2$. Prove that the Nottingham group, introduced in Exercise 4.3, is hereditarily just-infinite, i.e. that every open subgroup of the Nottingham group is just-infinite.

Exercise 9.5 (Saturable pro-p groups).

(a) Prove that every uniform pro-p group G satisfies $\gamma_p(G) \subseteq \Phi(G)^p$. Go on to show that, if G is a torsion-free finitely generated pro-p group satisfying $\gamma_p(G) \subseteq \Phi(G)^p$, then the lower p-series of G provides a potent filtration. Conclude from Theorem 8.4 that uniform pro-p groups are saturable.

(b) Let $d \in \mathbb{N}$ with $d \geq 3$, and let G be a Sylow pro-p subgroup of $\mathrm{GL}_d(\mathbb{Z}_p)$. Show that G is not uniform. (*Hint:* Show that the image of G in $\mathrm{GL}_d(\mathbb{F}_p)$ is not powerful.)

(c) Let $d \in \mathbb{N}$, and let G be the Sylow pro-p subgroup of $\mathrm{GL}_d(\mathbb{Z}_p)$. Determine the lower central series of G for the specific case $d = 3$ and guess the general pattern. (*Hint:* Take for G the group of matrices which are upper uni-triangular modulo p, and consider the commutators of elementary matrices.) Conclude from Theorem 8.4 that G is saturable for $d \leq p - 2$.

(d) Construct a torsion-free p-adic analytic pro-p group of dimension p which is not saturable. (*Hint:* Consider the semidirect product $G := A \ltimes M$ of the abelian groups $A = \langle \alpha \rangle \cong \mathbb{Z}_p$ and $M = \langle x_1, \ldots, x_{p-1} \rangle \cong \mathbb{Z}_p^{p-1}$, defined by

$$ x_i^\alpha = \begin{cases} x_i x_{i+1} & \text{if } 1 \leq i \leq p-2, \\ x_{p-1} x_1^p & \text{if } i = p-1. \end{cases} $$

Assume for a contradiction that G admits a potent filtration G_i, $i \in \mathbb{N}$. Observe that $[M,_{p-1} G] = M^p$ and deduce that $M \subseteq G_i$ for all $i \in \mathbb{N}$ in contradiction to $\bigcap\{G_i \mid i \in \mathbb{N}\} = 1$.)

Exercise 9.6 (Haar measure and random generation).
Every profinite group G is a compact topological group and as such it carries
a normalised Haar measure μ which is invariant under both left and right multiplication. The measure is normalised in the sense that $\mu(G) = 1$. The Haar
measure μ can be evaluated on Borel subsets, in particular on all closed subsets
of G. Sometimes it is useful to think of μ as a *probability measure* on G. For
$k \in \mathbb{N}$, it induces a probability measure on the direct product $G \times \ldots \times G$ of k
copies of the group G; thus one can consider *random k-tuples* of elements in G.
(a) Let G be profinite group and $H \leq_c G$. Determine the measure $\mu(H)$ in terms
of the index $|G : H|$.
(b) Let G be a finitely generated pro-p group, and put $d := d(G)$. For $k \in \mathbb{N}$,
determine the probability that a random k-tuple of elements in G generates G.
(c) Let G be a pro-p group of finite rank so that, by Exercise 6.6, its subgroup
growth is polynomially bounded: denoting by σ_n the number of subgroups of
index at most p^n in G, there exist $c, \alpha \in \mathbb{R}$ such that $\sigma_n \leq cp^{n\alpha}$ for all $n \in \mathbb{N}_0$.
Let $k \in \mathbb{N}$ with $k > \alpha + 1$. Deduce from the Borel–Cantelli lemma that a random
k-tuple of elements in G generates with probability 1 an open subgroup of G.
(*Hint:* The Borel–Cantelli lemma states that, if $X_i \subseteq_c G$, $i \in \mathbb{N}$, is a family of
closed subsets of G such that $\sum_{i=1}^{\infty} \mu(X_i)$ converges, then the Borel set

$$X = \bigcap \{Y_n \mid n \in \mathbb{N}\}, \quad \text{where } Y_n := \bigcup \{X_i \mid i \in \mathbb{N} \text{ with } i \geq n\},$$

has measure 0; see [44, Wind. 11]. In order to apply this in the given situation
note that a k-tuple fails to generate an open subgroup of G if and only if it is
contained in infinitely many open subgroups of G.)

Exercise 9.7 (Hausdorff dimension).
Let G be a pro-p group of finite rank, and write $G_n := G^{p^n}$ for $n \in \mathbb{N}$.
(a) Prove that

$$\dim(G) = \lim_{n \to \infty} \frac{\log_p |G : G_n|}{n}.$$

(*Hint:* Choose an open uniform subgroup U of G, and write $U_n := U^{p^n}$ for $n \in \mathbb{N}$.
Then $G_c \subseteq U$ for some $c \in \mathbb{N}$. Use the estimates $|U : U_{n-c}| \leq |G : G_n| \leq |G : U_n|$
for $n \in \mathbb{N}$ with $n \geq c$.)
(b) Suppose that G is uniform and let H be a uniform subgroup of G. Prove that
the *isolator* $\mathrm{iso}_G(H) := \{g \in G \mid \exists n \in \mathbb{N} : g^n \in H\}$ forms a uniform subgroup of
G with $|\mathrm{iso}_G(H) : H| < \infty$. (*Hint:* Work in the associated $\mathbb{Z}_p$-Lie lattice $L(G)$
and translate back and forth between the groups and the Lie lattices.)
(c) Let $H \leq_c G$ be a closed subgroup of G. Prove that

$$\lim_{n \to \infty} \frac{\log |HG_n : G_n|}{\log |G : G_n|} = \frac{\dim H}{\dim G}.$$

Remark: The limit on the left-hand side is equal to the *Hausdorff dimension* of
H in G with respect to (the metric induced by) the filtration G_n, $n \in \mathbb{N}$; see [3].

(*Hint:* Note that $|HG_n : G_n| = |H : H \cap G_n|$ for all $n \in \mathbb{N}$. Using similar arguments as in part (a), reduce to the situation where both G and H are uniform. Convince yourself that we can further assume that H is isolated in the group G, i.e. that $\mathrm{iso}_G(H) = H$. Employing the associated $\mathbb{Z}_p$-Lie lattices, prove the claim in this situation.)

Suggestions for further reading

Exercise 9.3 provides hands-on experience with p-adic quaternion groups; these groups provide interesting examples of anisotropic groups and appear in many contexts; e.g. see [25, 33, 34]. In connection with Exercise 9.4 we refer, once more, to [32], by Klaas, Leedham-Green and Plesken, as well as to Camina's and Wilson's chapters in [11]. As for Exercise 9.5, we remark that the connections and differences between uniformly powerful and saturable pro-p groups are clarified in [35]. The study of random generation in profinite groups, as discussed in Exercise 9.6, was initiated in Mann's stimulating paper [45]; also see [44, Wind. 11]. The notion of Hausdorff dimension provides a useful tool in the study of profinite groups. Exercise 9.7 was designed from results of Barnea and Shalev in their paper [3]. Other interesting applications can be found, for instance, in [2, 25].

10   Representations of compact p-adic Lie groups

In this section, we touch upon recent advances in the study of irreducible finite-dimensional complex linear representations of compact p-adic Lie groups; see [1]. Representation zeta functions are also discussed in Chapter III. Basic notions and facts on linear representations of groups can be found in Robinson's book [51, Ch. 8]. In [22], Huppert provides an accessible introduction to character theory.

10.1   Representation growth and Kirillov's orbit method

Let G be a profinite group. For $n \in \mathbb{N}$, we denote by $r_n(G)$ the number of isomorphism classes of continuous n-dimensional irreducible complex linear representations of G. Note that, by continuity, all representations of G of the kind considered have finite image. For a general profinite group G, the numbers $r_n(G)$ may very well be infinite, but there are interesting situations where they are all finite. Indeed, if G is finitely generated, then $r_n(G) < \infty$ for all $n \in \mathbb{N}$ if and only if G is FAb[5], i.e. if and only if $H/[H,H]$ is finite for every open subgroup $H \leq_o G$. (This can be proved from Jordan's theorem which states that, in characteristic 0, each finite linear group admits an abelian normal subgroup of index uniformly bounded in terms of the degree; cf. [44, Wind. 7].) In the situation where G is FAb one takes an interest in the arithmetic sequence $r_n(G)$, $n \in \mathbb{N}$,

---

[5]FAb sounds fabulous and is short for 'finite abelianisations'.

which reflects the *representation growth* of the profinite group G; cf. [37]. A useful tool is the *representation zeta function*

$$\zeta_G^{\mathrm{irr}}(s) := \sum_{n=1}^{\infty} r_n(G) n^{-s},$$

which encodes the entire representation growth of G.

The *derived series* of G is the descending series G_i, $i \in \mathbb{N}_0$, of closed normal subgroups, defined by $G_0 := G$ and $G_i := [G_{i-1}, G_{i-1}]$ for $i \geq 1$. It is easy to see that, if G is a pro-p group, then G is FAb if and only if every term of its derived series is open in G. In particular, any non-soluble just-infinite pro-p group is FAb.

Now suppose that G is p-adic analytic and consider the $\mathbb{Q}_p$-Lie algebra $\mathcal{L}(G) = \mathbb{Q}_p \otimes_{\mathbb{Z}_p} L(U)$ associated to G via an open uniform pro-p subgroup U; cf. Section 7.5. Then G is FAb if and only if $\mathcal{L}(G)$ is perfect, i.e. if and only if $\mathcal{L}(G) = [\mathcal{L}(G), \mathcal{L}(G)]$. This makes the representation growth of compact open subgroups of semisimple p-adic Lie groups a natural field of study, and the so-called orbit method provides a useful tool in this context.

Kirillov originally introduced the orbit method to study the unitary representations of nilpotent Lie groups in the 1960s, around the same time when Lazard developed his theory of p-adic Lie groups; see [30]. The method is based on the heuristic that there exists a close connection between the unitary irreducible representations of a Lie group and the orbits in its co-adjoint representation. In the case of a connected, simply connected nilpotent Lie group G with associated Lie algebra $\mathfrak{g}$, one obtains a natural correspondence between equivalence classes of irreducible unitary representations of G and G-orbits in the dual space of $\mathfrak{g}$. In [27], Kazhdan applied the orbit method to p-groups of nilpotency class less than p. In the late 1970s, Howe showed that the orbit method can also be put to use in the context of compact p-adic Lie groups; see [20]. More recently, in [24], Jaikin-Zapirain extended and applied the orbit method to solve problems in the subject of representation growth. González-Sánchez provides a summary in [14].

10.2   The orbit method for saturable pro-p groups

Let G be a saturable pro-p group and let $\mathfrak{g} := L(G)$ denote the associated $\mathbb{Z}_p$-Lie lattice. Continuous complex representations of G correspond to continuous complex characters. Thus we want to arrive at a description of the set $\mathrm{Irr}(G)$ of continuous irreducible complex characters of G.

In the following, we will frequently use the fact that the underlying sets of G and $\mathfrak{g}$ are one and the same. We denote by $\mathrm{Irr}(\mathfrak{g}_+)$ the set of continuous irreducible characters of the additive group $\mathfrak{g}_+ := (\mathfrak{g}, +)$, which coincides with the set $\mathrm{Hom}_{\mathbb{Z}}^{\mathrm{cont}}(\mathfrak{g}_+, \mathbb{C}^*)$ of continuous homomorphisms from $\mathfrak{g}_+$ into the multiplicative group $\mathbb{C}^*$. Indeed, one should think of $\mathrm{Irr}(\mathfrak{g}_+)$ as the dual space of $\mathfrak{g}$.

The adjoint action of G on $\mathfrak{g}$, which is given by conjugation, induces the so-called *co-adjoint action* of G on $\mathrm{Irr}(\mathfrak{g}_+)$: for $\omega \in \mathrm{Irr}(\mathfrak{g}_+)$ and $g \in G$, the element $\omega^g \in \mathrm{Irr}(\mathfrak{g}_+)$ is defined by setting

$$\omega^g(x) := \omega(x^{g^{-1}}) \quad \text{for all } x \in \mathfrak{g}.$$

Natural candidates for the irreducible characters of G arise in form of the class functions

$$\Phi_\Omega : G \to \mathbb{C}, \quad \Phi_\Omega(x) := |\Omega|^{-1/2} \sum_{\omega \in \Omega} \omega(x),$$

where Ω runs through all orbits of the G-space $\mathrm{Irr}(\mathfrak{g}_+)$. Indeed, one easily verifies that these functions form an orthonormal set. In fact, they give rise to an orthonormal basis for the class functions of G modulo any open PF-embedded subgroup N.

In parallel we need to keep track of the degrees of the irreducible characters of G. For this purpose, we introduce the notion of a radical. To $\omega \in \mathrm{Irr}(\mathfrak{g}_+)$ we associate the bi-additive and skew-symmetric map

$$b_\omega : \mathfrak{g} \times \mathfrak{g} \to \mathbb{C}^*, \quad b_\omega(x, y) := \omega([x, y]).$$

The *radical* of this map b_ω is

$$\mathrm{Rad}(\omega) := \mathrm{Rad}(b_\omega) = \{x \in \mathfrak{g} \mid \forall y \in \mathfrak{g} : b_\omega(x, y) = 1\}.$$

One can prove that the radical $\mathrm{Rad}(\omega)$ associated to $\omega \in \mathrm{Irr}(\mathfrak{g}_+)$ is, in fact, a Lie sublattice of $\mathfrak{g}$ and coincides as a set with a saturable subgroup of G, namely with the stabiliser $\mathrm{Stab}_G(\omega)$ of ω in G.

Theorem 10.1 (Orbit method for saturable pro-p groups). *Let G be a saturable pro-p group with $\gamma_{p-1}(G) \subseteq G^p$. Then the continuous irreducible complex characters of G are parameterised by the orbits of the co-adjoint action of G on $\mathrm{Irr}(\mathfrak{g}_+)$*

$$\mathrm{Irr}(G) = \{\Phi_\Omega \mid \Omega \text{ an orbit of the } G\text{-space } \mathrm{Irr}(\mathfrak{g}_+)\}.$$

Moreover, if Ω is the G-orbit of $\omega \in \mathrm{Irr}(\mathfrak{g}_+)$, then the degree of the corresponding irreducible character Φ_Ω is equal to $|\mathfrak{g} : \mathrm{Rad}(\omega)|^{1/2}$.

For a proof of this theorem, we refer to [14, §5]. Also we remark that the theorem applies in particular to uniform pro-p groups, if $p \geq 3$, and that technically more complicated conclusions hold true for uniform pro-2 groups.

Corollary 10.2. *Let G be a saturable pro-p group, which satisfies $\gamma_{p-1}(G) \subseteq G^p$ and which is FAb. Then*

$$\zeta_G^{\mathrm{irr}}(s) = \sum_{\omega \in \mathrm{Irr}(\mathfrak{g}_+)} |\mathfrak{g} : \mathrm{Rad}(\omega)|^{-1-s/2}.$$

Proof. Based on the theorem, this is now an easy computation. Observe that the dimension of the representation corresponding to a continuous complex character χ of G is equal to $\chi(1)$. Hence we have

$$
\begin{aligned}
\zeta_G^{\mathrm{irr}}(s) &= \sum_{\chi \in \mathrm{Irr}(G)} \chi(1)^{-s} \\
&= \sum_{\substack{\omega \in \mathrm{Irr}(\mathfrak{g}_+) \\ \Omega := \omega^G}} |\Omega|^{-1} \Phi_\Omega(1)^{-s} \\
&= \sum_{\omega \in \mathrm{Irr}(\mathfrak{g}_+)} |G : \mathrm{Stab}_G(\omega)|^{-1} |\mathfrak{g} : \mathrm{Rad}(\omega)|^{-s/2} \\
&= \sum_{\omega \in \mathrm{Irr}(\mathfrak{g}_+)} |\mathfrak{g} : \mathrm{Rad}(\omega)|^{-1-s/2}.
\end{aligned}
$$

$\square$

10.3 An application of the orbit method

Based upon Corollary 10.2, Jaikin-Zapirain uses in [24] model-theoretic techniques to show that, for odd primes p, the representation zeta function $\zeta_G^{\mathrm{irr}}(s) := \sum_{n=1}^{\infty} r_n(G) n^{-s}$ of a FAb p-adic analytic pro-p group G is in fact a rational function in p^{-s}. For $p = 2$, the same assertion holds if one further assumes that G is uniform. It is a major challenge to find out more about these representation zeta functions. Of particular interest are the zeta functions associated to families of open pro-p subgroups of semisimple p-adic Lie groups, such as the principal congruence subgroups of the special linear groups $\mathrm{SL}_n(\mathbb{Z}_p)$.

One of the main results announced in [1] establishes the existence of functional equations for families of pro-p groups derived from a global setting. For this, one considers families of p-adic Lie groups whose associated Lie algebras share a common $\mathbb{Z}$-Lie sublattice.

We describe a special case of the more general result stated in [1, Theorem A]. Denote by $\mathbb{P}$ the set of all primes. Let L be a Lie lattice over $\mathbb{Z}$, and for $p \in \mathbb{P}$ let $L_p := L \otimes_{\mathbb{Z}} \mathbb{Z}_p$ denote the completion of L at p. Then for all $p \in \mathbb{P}$ and $m \in \mathbb{N}$, with $m \geq 2$ if $p = 2$, the $\mathbb{Z}_p$-Lie lattice $p^m L_p$ is powerful and thus corresponds to a uniform pro-p group which we denote by $G_{p,m}$. A key example for this set-up is $L = \mathfrak{sl}_d(\mathbb{Z})$, $L_p = \mathfrak{sl}_d(\mathbb{Z}_p)$ and $G_{p,m} = \mathrm{SL}_d^m(\mathbb{Z}_p)$.

Theorem 10.3. *Let L be a Lie lattice over $\mathbb{Z}$ such that $\mathbb{Q} \otimes_{\mathbb{Z}} L$ is a perfect $\mathbb{Q}$-Lie algebra of dimension d. For $p \in \mathbb{P}$, consider the family of FAb uniform pro-p groups $G_{p,m}$ corresponding to the family of powerful $\mathbb{Z}_p$-Lie lattices $p^m L_p$, where $m \in \mathbb{N}$, with $m \geq 2$ if $p = 2$.*

Then for almost all $p \in \mathbb{P}$ the representation zeta functions associated to the groups $G_{p,m}$, m as above, satisfy the functional equations

$$
\zeta_{G_{p,m}}^{\mathrm{irr}}(s)|_{p \to p^{-1}} = p^{(1-2m)d} \zeta_{G_{p,m}}^{\mathrm{irr}}(s).
$$

These functional equations are to be interpreted as follows. Consider $G_{p,m}$ for $p \in \mathbb{P}$ and $m \in \mathbb{N}$ as above. The zeta function $\zeta_{G_{p,m}}^{\mathrm{irr}}(s)$ is a rational function in p^{-s} whose coefficients can be expressed as polynomials in p and in the numbers $\nu(V)$ of $\mathbb{F}_p$-points of certain smooth projective $\mathbb{F}_p$-defined varieties V. The operation $p \to p^{-1}$ on such a number $\nu(V)$ is performed by inverting the Frobenius eigenvalues associated to V: the alternating sum of these complex numbers equals $\nu(V)$ in accordance with the Weil conjectures. In the simplest case, the Frobenius eigenvalues are just powers of p, in agreement with our notation.

The proof of Theorem 10.3 is built on two main ideas. The first ingredient is the parameterisation of the irreducible characters of a FAb uniform pro-p group G in terms of the orbits of the co-adjoint action of G. In a second step, one takes advantage of the fact that the problem of counting co-adjoint orbits can be treated within the framework of generalised Igusa local zeta functions; cf. Chapter III and the references therein.

Suggestions for further reading

The orbit method in a wide context is discussed in Kirillov's book [31]. The 'classical' articles, by Kirillov, Kazhdan and Howe, referred to in the main text are [30, 27, 20]. More recent articles which develop or apply the orbit method in the context of representation zeta functions include [24, 7, 14, 37, 1]. They lead the reader to cutting-edge research in asymptotic group theory.

References for Chapter I

[1] N. Avni, B. Klopsch, U. Onn and C. Voll. On representation zeta functions of groups and a conjecture of Larsen–Lubotzky. *C. R. Math. Acad. Sci. Paris*, 348(7-8): 363–367, 2010.

[2] Y. Barnea and B. Klopsch. Index-subgroups of the Nottingham group. *Adv. Math.*, 180(1): 187–221, 2003.

[3] Y. Barnea and A. Shalev. Hausdorff dimension, pro-p groups, and Kac–Moody algebras. *Trans. Amer. Math. Soc.*, 349(12): 5073–5091, 1997.

[4] S. R. Blackburn, P. M. Neumann and G. Venkataraman. *Enumeration of finite groups*, volume 173 of *Cambridge Tracts in Mathematics*. Cambridge University Press, Cambridge, 2007.

[5] N. Bourbaki. *General topology: Chapters 1–4. Elements of Mathematics*. Springer-Verlag, Berlin, 1998.

[6] N. Bourbaki. *Lie groups and Lie algebras: Chapters 1–3. Elements of Mathematics*. Springer-Verlag, Berlin, 1998.

[7] M. Boyarchenko and M. Sabitova. The orbit method for profinite groups and a p-adic analogue of Brown's theorem. *Israel J. Math.*, 165: 67–91, 2008.

[8]   D. Bump, J. W. Cogdell, E. de Shalit, D. Gaitsgory, E. Kowalski and
      S. S. Kudla. *An introduction to the Langlands program.* Birkhäuser Boston
      Inc., Boston, MA, 2003. Lectures presented at the Hebrew University of
      Jerusalem, Jerusalem, 12–16 March 2001, Edited by Joseph Bernstein and
      Stephen Gelbart.

[9]   C. Chevalley. *Theory of Lie groups. I,* volume 8 of *Princeton Mathematical
      Series.* Princeton University Press, Princeton, NJ, 1999.

[10] J. D. Dixon, M. P. F. du Sautoy, A. Mann and D. Segal. *Analytic pro-
      p groups,* volume 61 of *Cambridge Studies in Advanced Mathematics.*
      Cambridge University Press, Cambridge, second edition, 1999.

[11] M. P. F. du Sautoy, D. Segal and A. Shalev, editors. *New horizons in pro-p
      groups,* volume 184 of *Progress in Mathematics.* Birkhäuser Boston Inc.,
      Boston, MA, 2000.

[12] H.-D. Ebbinghaus, H. Hermes, F. Hirzebruch, M. Koecher, K. Mainzer,
      J. Neukirch, A. Prestel and R. Remmert. *Numbers,* volume 123 of *Graduate
      Texts in Mathematics.* Springer-Verlag, New York, 1991.

[13] J. González-Sánchez. On p-saturable groups. *J. Algebra,* 315(2): 809–823,
      2007.

[14] J. González-Sánchez. Kirillov's orbit method for p-groups and pro-p groups.
      Comm. Algebra, 37(12): 4476–4488, 2009.

[15] J. González-Sánchez and Benjamin Klopsch. Analytic pro-p groups of small
      dimensions. *J. Group Theory,* 12(5): 711–734, 2009.

[16] F. Q. Gouvêa. *p-adic numbers. An introduction.* Universitext. Springer-
      Verlag, Berlin, second edition, 1997.

[17] P. Hall. *The Edmonton notes on nilpotent groups.* Queen Mary College
      Mathematics Notes. Mathematics Department, Queen Mary College,
      London, 1969.

[18] P. J. Higgins. *Introduction to topological groups,* volume 15 of *London
      Mathematical Society Lecture Note Series.* Cambridge University Press,
      London, 1974.

[19] R. Hooke. Linear p-adic groups and their Lie algebras. *Ann. of Math. (2),*
      43: 641–655, 1942.

[20] R. E. Howe. Kirillov theory for compact p-adic groups. *Pacific J. Math.,*
      73(2): 365–381, 1977.

[21] B. Huppert. *Endliche Gruppen. I. Die Grundlehren der Mathematischen
      Wissenschaften,* Band 134. Springer-Verlag, Berlin, 1967.

[22] B. Huppert. *Character theory of finite groups*, volume 25 of *de Gruyter Expositions in Mathematics*. Walter de Gruyter & Co., Berlin, 1998.

[23] I. Ilani. Analytic pro-p groups and their Lie algebras. *J. Algebra*, 176(1): 34–58, 1995.

[24] A. Jaikin-Zapirain. Zeta function of representations of compact p-adic analytic groups. *J. Amer. Math. Soc.*, 19(1): 91–118, 2006.

[25] A. Jaikin-Zapirain and B. Klopsch. Analytic groups over general pro-p domains. *J. Lond. Math. Soc. (2)*, 76(2): 365–383, 2007.

[26] U. Jannsen and K. Wingberg. Die Struktur der absoluten Galoisgruppe p-adischer Zahlkörper. *Invent. Math.*, 70(1): 71–98, 1982/83.

[27] D. Kazhdan. Proof of Springer's hypothesis. *Israel J. Math.*, 28(4): 272–286, 1977.

[28] J. L. Kelley. *General topology*, volume 27 of *Graduate Texts in Mathematics*. Springer-Verlag, New York, 1975.

[29] E. I. Khukhro. *p-automorphisms of finite p-groups*, volume 246 of *London Mathematical Society Lecture Note Series*. Cambridge University Press, Cambridge, 1998.

[30] A. A. Kirillov. Unitary representations of nilpotent Lie groups. *Uspehi Mat. Nauk*, 17(4 (106)): 57–110, 1962.

[31] A. A. Kirillov. *Lectures on the orbit method*, volume 64 of *Graduate Studies in Mathematics*. American Mathematical Society, Providence, RI, 2004.

[32] G. Klaas, C. R. Leedham-Green and W. Plesken. *Linear pro-p-groups of finite width*, volume 1674 of *Lecture Notes in Mathematics*. Springer-Verlag, Berlin, 1997.

[33] B. Klopsch. Pro-p groups with linear subgroup growth. *Math. Z.*, 245(2): 335–370, 2003.

[34] B. Klopsch. Groups with less than n subgroups of index n. *Math. Ann.*, 333(1): 67–85, 2005.

[35] B. Klopsch. On the Lie theory of p-adic analytic groups. *Math. Z.*, 249(4): 713–730, 2005.

[36] Y. Kosmann-Schwarzbach. *Groups and symmetries. From finite groups to Lie groups*. Universitext. Springer, New York, 2010.

[37] M. Larsen and A. Lubotzky. Representation growth of linear groups. *J. Eur. Math. Soc. (JEMS)*, 10(2): 351–390, 2008.

[38] M. Lazard. Sur les groupes nilpotents et les anneaux de Lie. *Ann. Sci. Ecole Norm. Sup. (3)*, 71: 101–190, 1954.

[39] M. Lazard. Groupes analytiques p-adiques. *Inst. Hautes Études Sci. Publ. Math.* 26: 389–603, 1965.

[40] C. R. Leedham-Green and S. McKay. *The structure of groups of prime power order*, volume 27 of *London Mathematical Society Monographs. New Series*. Oxford University Press, Oxford, 2002.

[41] M. W. Liebeck and A. Shalev. Diameters of finite simple groups: sharp bounds and applications. *Ann. of Math. (2)*, 154(2): 383–406, 2001.

[42] A. Lubotzky and A. Mann. Powerful p-groups. I. Finite groups. *J. Algebra*, 105(2): 484–505, 1987.

[43] A. Lubotzky and A. Mann. Powerful p-groups. II. p-adic analytic groups. *J. Algebra*, 105(2): 506–515, 1987.

[44] A. Lubotzky and D. Segal. *Subgroup growth*, volume 212 of *Progress in Mathematics*. Birkhäuser Verlag, Basel, 2003.

[45] A. Mann. Positively finitely generated groups. *Forum Math.*, 8(4): 429–459, 1996.

[46] G. Navarro. The McKay conjecture and Galois automorphisms. *Ann. of Math. (2)*, 160(3): 1129–1140, 2004.

[47] N. Nikolov and D. Segal. On finitely generated profinite groups. I. Strong completeness and uniform bounds. *Ann. of Math. (2)*, 165(1): 171–238, 2007.

[48] E. A. O'Brien and M. R. Vaughan-Lee. The groups with order p^7 for odd prime p. *J. Algebra*, 292(1): 243–258, 2005.

[49] C. Reutenauer. *Free Lie algebras*, volume 7 of *London Mathematical Society Monographs. New Series*. The Clarendon Press, Oxford University Press, New York, 1993.

[50] L. Ribes and P. Zalesskii. *Profinite groups*, volume 40 of *Ergebnisse der Mathematik und ihrer Grenzgebiete. 3. Folge*. Springer-Verlag, Berlin, 2000.

[51] D. J. S. Robinson. *A course in the theory of groups*, volume 80 of *Graduate Texts in Mathematics*. Springer-Verlag, New York, second edition, 1996.

[52] W. Rossmann. *Lie groups. An introduction through linear groups*, volume 5 of *Oxford Graduate Texts in Mathematics*. Oxford University Press, Oxford, 2002.

[53] D. Segal. *Words: notes on verbal width in groups*, volume 361 of *London Mathematical Society Lecture Note Series*. Cambridge University Press, Cambridge, 2009.

[54] J.-P. Serre. *Galois cohomology. Springer Monographs in Mathematics.* Springer-Verlag, Berlin, English edition, 2002.

[55] J.-P. Serre. *Lie algebras and Lie groups*, volume 1500 of *Lecture Notes in Mathematics*. Springer-Verlag, Berlin, 2006.

[56] A. Shalev. Characterization of p-adic analytic groups in terms of wreath products. *J. Algebra*, 145(1): 204–208, 1992.

[57] A. Shalev. Asymptotic group theory. *Notices Amer. Math. Soc.*, 48(4): 383–389, 2001.

[58] A. Weil. *Basic number theory. Classics in Mathematics.* Springer-Verlag, Berlin, 1995.

[59] S. Willard. *General topology.* Dover Publications Inc., Mineola, NY, 2004.

[60] J. S. Wilson. *Profinite groups*, volume 19 of *London Mathematical Society Monographs. New Series.* The Clarendon Press, Oxford University Press, New York, 1998.

Chapter II

Strong approximation methods
by Nikolay Nikolov

1 Introduction

This chapter is concerned with linear groups $\Gamma \leq \mathrm{GL}_n(k)$ where k is some field (usually of characteristic 0). Linearity is one of the most effective and well-studied conditions one can put on a general infinite group. The following three results are some of the most widely used consequences of linearity:

(i) **(Mal'cev):** A finitely generated linear group Γ is residually finite.

(ii) **(Selberg):** If in addition char $k = 0$, then Γ is virtually torsion free.

(iii) **(Tits alternative):** A finitely generated linear group Γ is either virtually soluble or else contains a non-abelian free group.

The first of these means that Γ has many finite images, and one way to study Γ is to investigate these images (equivalently, the profinite completion $\widehat{\Gamma}$ of Γ). In turn, the Tits alternative (iii) can be interpreted as saying that a non-virtually soluble linear group Γ is rather big and a natural question is whether this implies some extra properties of $\widehat{\Gamma}$. In other words, if G is not virtually soluble, then does Γ have (in some sense) a rich collection of finite images?

This is indeed true and the answer is provided by the following 'Lubotzky alternative' for linear groups, one of the main objectives of this chapter:

Theorem 1.1. *Let $\Delta \leq \mathrm{GL}_n(k)$ be a finitely generated linear group over a field k of characteristic 0. Then one of the following holds:*

(a) *the group Δ is virtually soluble, or*

(b) *there exists a connected, simply connected $\mathbb{Q}$-simple algebraic group G, a finite set of primes S such that $\Gamma = G(\mathbb{Z}_S)$ is infinite and a subgroup Δ_1 of finite index in Δ such that every congruence image of Γ appears as a quotient of Δ_1.*

Here $\mathbb{Z}_S = \mathbb{Z}[1/p \mid p \in S]$. In case (b) we can deduce from the *strong approximation theorem* that Δ_1 has many finite images, in particular the groups $\prod_{i=1}^k G(\mathbb{F}_{p_i})$ whenever $p_1, \ldots, p_k$ are distinct primes outside S. Now, for all but finitely many primes p, the group $G(\mathbb{F}_p)$ is semisimple, in fact it is a perfect central extension of a product of simple groups (of fixed Lie type over $\mathbb{F}_p$). The simple groups of Lie type are very well understood and this enables us to deduce properties of the profinite completion $\widehat{\Delta}$ of Δ.

For example, if Δ has polynomial subgroup growth, then one can deduce that case (b) of Theorem 1.1 is impossible and hence that Δ is virtually soluble. Some more applications of Theorem 1.1 are given in Section 6 below.

In turn, when Δ is virtually soluble we have the **Lie–Kolchin theorem**:

Theorem 1.2. *Suppose that $\Delta \leq \mathrm{GL}_n(K)$ is a virtually soluble linear group over an algebraically closed field K. Then Δ has a triangularisable subgroup Δ_1 of finite index; i.e. Δ_1 is conjugate to a subgroup of the upper triangular matrices in $\mathrm{GL}_n(K)$.*

In fact, if char $K = 0$, the index of Δ_1 in Δ can be bounded by a function of n only (a theorem of Mal'cev and Platonov). This has the corollary:

Lemma 1.3. *Suppose that Δ is a finitely generated group which is residually in the class of virtually soluble linear groups of degree n in characteristic 0. Then Δ itself is virtually soluble.*

We shall use this Lemma in the proof of Theorem 1.1.

A common feature in the proofs of all these results is to take the *Zariski closure* $G = \overline{\Delta}$ of Δ in $\mathrm{GL}_n(K)$. This is a linear algebraic group and we can apply results from algebraic geometry, number theory and the theory of arithmetic groups to study G and its dense subgroup Δ.

The main object of this chapter is to understand the terminology appearing above and develop the methods by which Theorems 1.1 and 1.2 can be proved. These methods are useful in a variety of other situations involving linear groups, some of which are discussed in Section 6 below.

2   Algebraic groups

Throughout, K will denote an algebraically closed field.

2.1   The Zariski topology on K^n

A good reference for the material of this section (with proofs) is the book [1] by Atiyah and Macdonald. For a brief introduction, see also the chapter 'Linear algebraic groups' in [5].

Let K^n be the n-dimensional vector space over K. Given a subset S of the polynomial ring

$$R := K[x_1, \ldots, x_n]$$

define
$$V(S) = \{x \in K^n \mid f(\mathbf{x}) = 0 \quad \forall f \in S\}$$
to be set of common zeros of S in K^n.

It is easy to see that $V(I) = V(S)$ for the ideal I generated by S, that

$$V(I) \cup V(J) = V(IJ)$$

for ideals I and J of R, and that

$$\bigcap_{I \in \mathcal{F}} V(I) = V\left(\sum_{I \in \mathcal{F}} I\right)$$

for any family $\mathcal{F}$ of ideals of R.

The Hilbert basis theorem says that each ideal I of R is finitely generated and so each $V(S)$ can in fact be defined by finitely many polynomial equations.

Definition 2.1. The *Zariski topology* of K^n has as its closed sets the sets $V(I)$ for all ideals I of R.

The first basic result about the Zariski topology is the following:

Proposition 2.2 (Exercise 8.3). *The space K^n with the Zariski topology is a compact topological space, in fact it satisfies the descending chain condition on closed subsets.*

Note that the closed sets of K coincide with its finite subsets (since a polynomial in one variable can only have finitely many roots). More generally, the Zariski topology of K^n is never Hausdorff, thus even though the space K^n is compact one should be careful when applying familiar results from Hausdorff spaces.

Example: Let V be the hyperbola given by the equation $x_1 x_2 = 1$ in K^2. Then V is a closed, hence compact subset of K^2, but its projection on the x_1 axis is $K \backslash \{0\}$, which is not closed. So in the Zariski topology, continuous images of compact sets are not always closed.

The subsets $V(I) \subseteq K^n$ (with the subspace topology induced from the Zariski topology on K^n) are called *affine (algebraic) varieties*. If W is an affine variety, the *coordinate ring* $R(W)$ of W is the algebra $R/J(W)$, where $J(W)$ is the ideal of R consisting of all polynomials vanishing on W. The ascending chain condition on ideals of R (Hilbert's basis theorem) implies the descending chain condition (minimal condition) for closed sets in K^n.

Theorem 2.3 (Hilbert's Nullstellensatz). *$V(I) = \emptyset$ if and only if $I = R$. More generally, if $W = V(I)$ is an affine variety, then $J(W)/I$ is the radical of I, i.e.*

$$J(W) = \{x \in R \mid x^n \in I \text{ for some } n \in \mathbb{N}\}.$$

For a proof, see [1, Ch. 7].

The coordinate ring $R(W)$ can be considered as the set of morphisms of W into the 1-dimensional variety K. In general, a *morphism* F from $W_1 \subseteq K^{n_1}$ into $W_2 \subseteq K^{n_2}$ is an n_2-tuple $(f_1, \ldots, f_{n_2}) \in K[x_1, \ldots, x_{n_1}]^{n_2}$ of polynomial maps such that $F(W_1) \subseteq W_2$. Any such morphism induces a K-algebra homomorphism $F^* : R(W_2) \to R(W_1)$ defined by $f \mapsto f \circ F$. Conversely, from the Nullstellensatz it can be shown that every algebra homomorphism F^* between $R(V_2)$ and $R(V_1)$ arises in this way from a morphism $F : V_1 \to V_2$. In this way the category of affine varieties is anti-equivalent to the category of reduced[1] finitely generated algebras over the algebraically closed field K.

Definition 2.4. A variety is *irreducible* if it is not the union of two proper closed subsets.

Since V satisfies the minimal condition on closed subsets, we can write every variety W as
$$W = W_1 \cup W_2 \cup \cdots \cup W_k,$$
a union of irreducible varieties W_i. If we assume that the above decomposition is irredundant, i.e. $W_i \not\subseteq W_j$ whenever $i \neq j$, then it is in fact unique up to reordering of the W_i. These are then called the *irreducible components* of W.

For example, if W is the variety defined by the single equation
$$x_1 x_2 \left(x_1 x_2^2 - 1 \right) = 0,$$
then its irreducible components are the two lines with equations $x_1 = 0$, $x_2 = 0$ and the curve defined by $x_1 x_2^2 = 1$.

It is easy to see that a variety W is irreducible if and only if $J(W)$ is a prime ideal of W, i.e. if and only if the coordinate ring $R/J(W)$ is an integral domain.

Definition 2.5. The *dimension*, $\dim W$ of an irreducible variety W is the Krull dimension of $R(W)$. This is the transcendence degree of $R(W)$ over K, or equivalently the maximal length d of a chain of distinct non-zero prime ideals $0 \subset P_1 \subset \cdots \subset P_d \subset R(W)$. The dimension of a general affine variety is the maximal dimension of its irreducible components.

As a consequence, a closed proper subset of an irreducible variety W has strictly smaller dimension than W.

2.2 Linear algebraic groups as closed subgroups of $\mathrm{GL}_n(K)$

We identify the $n \times n$ matrix ring $M_n(K)$ with K^{n^2}, and use x_{ij} $(i, j = 1, \ldots, n)$ as coordinates. Then the subgroup $\mathrm{SL}_n(K)$ of matrices with determinant 1 forms an affine variety, defined by the equation $\det(x_{ij}) = 1$.

Definition 2.6. A *linear algebraic group* over K is a Zariski-closed subgroup of $\mathrm{SL}_n(K)$ for some n.

---

[1]A ring or an algebra is *reduced* if it contains no nilpotent elements except 0.

Notes:

1. The two maps $(x, y) \mapsto xy$ and $x \mapsto x^{-1}$ from $G \times G$ (respectively G) to G are morphisms of affine varieties.

2. There are more general algebraic groups which are not linear. In this book we shall be concerned only with linear algebraic groups, and 'algebraic group' will always mean 'linear algebraic group'.

3. The definition we have given is different from the standard one but equivalent to it: one usually defines a linear algebraic group to be an affine variety with maps of group multiplication and inverses which are morphisms of varieties. It can be shown that every such group is in fact isomorphic to a closed subset of some $SL_n(K)$. See the 'Linear algebraic groups' chapter in [5].

A *homomorphism* of linear algebraic groups $f : G \to H$ is a group homomorphism which is also a morphism between varieties, i.e. f is given by polynomial maps on the realisations of $G \subset M_{n_1}(K)$ and $H \subset M_{n_2}(K)$.

The group $GL_n(K)$ is isomorphic to a closed subgroup of $SL_{m+1}(K)$, by the mapping

$$g \mapsto \begin{pmatrix} g & 0 \\ 0 & (\det g)^{-1} \end{pmatrix}.$$

In this way we consider $GL_n(K)$ as a linear algebraic group. It is clear that every linear algebraic group is isomorphic to a closed subgroup of $GL_n(K)$ for some n.

Basic examples

For an integer $n \geq 2$, consider the following subgroups of $SL_n(K)$:

- the group of (upper or lower) unitriangular matrices,

- the upper (upper or lower) triangular matrices,

- the diagonal matrices, or more generally,

- the monomial matrices.

It is clear that these are closed subgroups of $SL_n(K)$ and so are algebraic groups.

Note that when $n = 2$ the first example is isomorphic to the additive group of the field K, while the third one is isomorphic to the multiplicative group of K. In this way $(K, +)$ and $(K, \times)$ become linear algebraic groups. The first one is denoted by $\mathbb{G}_+$ and the second by $\mathbb{G}_\times$. In can be shown that these are the only connected algebraic groups of dimension 1.

Another family of examples arises from linear groups preserving some form. For example, if $(\mathbf{u}, \mathbf{v}) = \mathbf{u}^T P \mathbf{v}$ is a bilinear form on the vector space $V = K^n$, then the group $G \leq \mathrm{GL}(V)$ preserving $(-, -)$ can be described as those matrices X in $\mathrm{GL}_n(K)$ such that $X^T P X = P$. This is a collection of n^2 polynomial equations on the coefficients of $X = (x_{ij})$ and so G is an algebraic group. Examples are the symplectic group $\mathrm{Sp}_n(K)$ (n even) and the special orthogonal group $\mathrm{SO}_n(K)$.

Basic properties of Algebraic groups

Theorem 2.7 (see [7, Ch. II]). *Let $f : G \to H$ be a homomorphism of algebraic groups. Then:*

1. $\mathrm{Im}(f)$ *is a closed subgroup of H and* $\ker(f)$ *is a closed subgroup of G.*

2. $\dim G = \dim \ker(f) + \dim \mathrm{Im}(f)$.

Recall that a topological space is *connected* if it cannot be written as the disjoint union of two proper closed (equivalently open) subsets. Clearly, an irreducible variety is connected. It turns out that for algebraic groups the converse is also true and so the two concepts coincide. To see this, suppose that G is a connected algebraic group. Let $G = V_1 \cup \cdots \cup V_k$ be the decomposition of G into irreducible components. This decomposition is unique up to the order of the V_i, therefore the action of G by left multiplication permutes the components V_i. Without loss of generality, suppose that $1 \in V_1$. Let

$$G_1 = \mathrm{Stab}_G(V_1) := \{g \in G \mid gV_1 = V_1\}.$$

Clearly, G_1 is a closed subgroup of finite index k in G, so it is both open and closed, as are each of its cosets in G. Since G is connected, we must have $G = G_1$, so $k = 1$ and G is irreducible.

The above argument easily shows that more generally the connected component of the identity G° of G is a closed irreducible normal subgroup of finite index in G; it may be characterised as the smallest closed subgroup of finite index in G.

Lemma 2.8 (see [20, 14.15], or [7, §7.5]). *If $(H_i)_{i \in I}$ is a family of closed connected subgroups of G, then the subgroup $\langle H_i | \ i \in I \rangle$ generated abstractly by the H_i in G is closed and connected.*

In particular, if H_1 and H_2 are two closed connected subgroups of G such that $H_1 H_2 = H_2 H_1$ (e.g., if either of H_1 or H_2 is normal in G), then $H_1 H_2$ is a closed connected subgroup of G. In general, if H_1, H_2 are closed subgroups with $H_1 H_2 = H_2 H_1$ and having connected components H_1^0, H_2^0 respectively, then $H_1 H_2$ is a finite union of closed sets $h H_1^0 H_2^0 h'$ for some $h, h' \in G$ and so is a closed subgroup of G.

Theorem 2.9 (Chevalley; see [7, Ch. IV]). *If H is a closed normal subgroup of G, then the quotient G/H can be given the structure of a linear algebraic group, so that the quotient map $G \to G/H$ is a homomorphism of algebraic groups.*

Fields of definition and restriction of scalars

A variety $V(I)$ is said to be *defined over* a subfield $k \subset K$ if the ideal I is generated (as an ideal of R) by polynomials with coefficients in k. When the field extension K/k is separable (which is always the case if k has characteristic 0) there is a useful criterion for V to be defined over k:

Lemma 2.10. *Let $W = V(S)$ be a variety. For $\sigma \in \mathrm{Gal}(K/k)$, define the variety W^σ to be $V(S^\sigma)$, i.e. the zero set of the ideal S^σ of R. Then W is defined over k if and only if $W = W^\sigma$ for all $\sigma \in \mathrm{Gal}(K/k)$.*

Similarly, a homomorphism $f : G \to H$ between two algebraic groups is k-defined if all the coordinate maps defining f are polynomials with entries in k.

Now let $G \leq \mathrm{GL}_n(K)$ be an algebraic group and let $\mathcal{O}$ be a subring of K. The group of $\mathcal{O}$-*rational points* of G is defined to be $\mathrm{GL}_n(\mathcal{O}) \cap G$ and is denoted by $G_\mathcal{O}$.

Suppose that G is defined over some subfield k of K which is a finite extension of k_0. In this chapter we shall study the groups G_k and sometimes we prefer to reduce the situation to a smaller subfield k_0 (which will usually be $\mathbb{Q}$).

There is a standard procedure for doing this, called 'restriction of scalars'. This associates to G another algebraic group $H \leq \mathrm{GL}_{nd}(K)$ where $d = (k : k_0)$; here H is defined over k_0 and satisfies $H_{k_0} = G_k$. The algebraic group H is denoted $\mathcal{R}_{k/k_0}(G)$. Before presenting the general construction let us study a simple special case which illustrates the idea.

Suppose that G is the multiplicative group of the field $(K, \times)$. This is defined over the integers $\mathbb{Z}$ (i.e. it can be defined by polynomials over $\mathbb{Z}$.) Let k be a number field, i.e. a finite extension field of $\mathbb{Q}$. The group G_k is clearly the multiplicative group k^* of the field k. We want to find a $\mathbb{Q}$-defined algebraic group H such that its group $H_\mathbb{Q}$ of $\mathbb{Q}$-rational points is isomorphic (as an abstract group) to G_k.

To find H we identify k with the vector space $\mathbb{Q}^d$ by choosing a basis $a_1, \ldots, a_d$ for k over $\mathbb{Q}$, and consider the regular representation of k acting on itself by left multiplication. This gives an algebra monomorphism $\rho : k \to M_d(\mathbb{Q})$ and so $\rho(k)$ is a d-dimensional subspace of $M_n(\mathbb{Q})$. This can be defined as the zeros of some $s = d^2 - d$ linear functionals $F_1, \ldots, F_s : M_n(\mathbb{Q}) \to \mathbb{Q}$. Therefore, we can define the algebraic variety H as the set of zeros of $F_1, \ldots, F_s$ in $\mathrm{GL}_d(K)$. Then clearly $H_\mathbb{Q} = G_k$ and the only thing that has to be checked is that H is a group, i.e. the variety H is closed under matrix multiplication and inverses. This can be expressed as the vanishing of certain polynomials in the coordinates x_{ij}. If one of these polynomials is non-trivial, it will be non-trivial for some rational values of its arguments. But we certainly know that $H_\mathbb{Q}$ is closed under multiplication and inverses since it is equal to the multiplicative group k^*. So H is indeed an algebraic group.

There is another way to view the algebraic group H just constructed. Let $\sigma_1, \ldots, \sigma_d$ be the d embeddings of k into the algebraically closed field K. For an element $h = \sum_{i=1}^{d} x_i a_i \in k$ with $x_i \in \mathbb{Q}$, consider

$$\lambda(h) = (\lambda_1(h), \ldots, \lambda_d(h)),$$

where

$$\lambda_j(h) = \sum_{i=1}^{d} x_i \sigma_j(a_i) = \sigma_j(h).$$

The condition $\det \rho(h) \neq 0$ is equivalent to $\prod_j \lambda_j(h) \neq 0$. If $k = \mathbb{Q}(\alpha_1)$ where α_1 has minimal polynomial $p(x) = (x - \alpha_1) \cdots (x - \alpha_d)$ over $\mathbb{Q}$, then $k \simeq \mathbb{Q}[x]/(p(x))$. We can extend λ from k to $k \otimes_{\mathbb{Q}} \mathbb{C}$ and then

$$k \otimes_{\mathbb{Q}} \mathbb{C} \simeq \frac{\mathbb{C}[x]}{(p(x))} \simeq \bigoplus_{i=1}^{d} \frac{\mathbb{C}[x]}{(x - \alpha_i)}, \tag{2.1}$$

where the second isomorphism comes from the Chinese remainder theorem and coincides with λ. Thus $\lambda \circ \rho^{-1}$ provides a K-isomorphism of H with the direct product $(\mathbb{G}_{\times})^d$ of d copies of the multiplicative group $\mathbb{G}_{\times}$.

In general, we are given a k-defined algebraic group $G \leq \mathrm{GL}_n(K)$. Consider again an embedding $\rho : k \to M_d(k_0)$ given by the regular representation of k acting on itself. Again the subspace $\rho(k) \subset M_d(k_0)$ is defined by some set of say r linear equations $F_i(y_{ab})$ in the matrix entries y_{ab} ($1 \leq a, b \leq d$ and $1 \leq i \leq r$). Suppose that G was defined as a variety by the l polynomials $P_j(z^{st})$ in the entries of the matrix $(z^{st}) \in M_n(K)$ ($j = 1, \ldots, l$, $1 \leq s, t \leq n$).

Now the algebraic group $H = \mathcal{R}_{k/k_0}(G)$ is defined by the following two families of equations in the $(nd)^2$ variables z_{ab}^{st}:

The first family is

$$P_j \left(\left(z_{ab}^{st} \right)_{a,b} \right) = 0 \in M_d(K), \quad j = 1, 2, \ldots, l,$$

i.e. we replace each variable z^{st} in the original polynomial P_j with a matrix $(z_{ab}^{st})_{a,b} \in M_d(K)$. Note that each P_j gives d^2 polynomial equations in K, one for each entry of the matrix in $M_d(K)$.

The second family is

$$F_i \left(\left(z_{ab}^{st} \right)_{a,b} \right) = 0, \quad i = 1, \ldots, r,$$

for each pair (s, t) with $1 \leq s, t \leq n$.

A typical example is the group

$$G = \left\{ \begin{pmatrix} a & 2b \\ b & a \end{pmatrix} \ \middle| \ a^2 - 2b^2 \neq 0 \right\}$$

which is the restriction of scalars $\mathcal{R}_{\mathbb{Q}(\sqrt{2})/\mathbb{Q}} \mathbb{G}_{\times}$. Here, G is K-isomorphic to $\mathbb{G}_{\times} \times \mathbb{G}_{\times}$ via the map $\begin{pmatrix} a & 2b \\ b & a \end{pmatrix} \mapsto (a + ib, a - ib)$, but this isomorphism is not $\mathbb{Q}$-defined.

It is easy to see that if we have a k-defined morphism $f : G \to T$ between two k-defined linear algebraic groups, then this induces a k_0-defined morphism

$$\mathcal{R}_{k/k_0}(f) : \mathcal{R}_{k/k_0}(G) \to \mathcal{R}_{k/k_0}(T).$$

In this way $\mathcal{R}_{k/k_0}$ becomes a functor between the category of k-defined groups and morphisms and k_0-defined groups and morphisms.

The Lie algebra of G

There is a standard way to associate a Lie algebra $L(G)$ to any connected linear algebraic group G, so that the map $L : G \mapsto L(G)$ is an equivalence of categories. More precisely, the following holds (see III of [7]):

- If $f : G \to H$ is a homomorphism of algebraic groups, then there is a uniquely specified homomorphism $L(f) : L(G) \to L(H)$ between their Lie algebras.

- In particular, for any given $g \in G$ the map $x \mapsto g^{-1}xg$ is an automorphism of G and this gives rise to a Lie algebra automorphism denoted $\mathrm{Ad}g :$ $L(G) \to L(G)$. In this way we get a homomorphism of algebraic groups $\mathrm{Ad} : G \to \mathrm{Aut}L(G)$, and it is easy to see that $\ker \mathrm{Ad} = Z(G)$, the centre of G. This is the adjoint action of G on $L(G)$.

- If H is a closed (normal) subgroup of G, then $L(H)$ is a Lie subalgebra (respectively an ideal) of $L(G)$.

- If G is defined over a subfield k of K, then $L(G)$ is also defined over k, i.e. it has a basis such that the structure constants of the Lie bracket multiplication are elements of k. Moreover, if the morphism $f : G \to H$ is k-defined, then so is the Lie algebra homomorphism $L(f)$.

- The dimension of $L(G)$ (as a vector space over K) is equal to the dimension of the algebraic group G.

In general, if G is not connected, we define $L(G)$ to be equal to $L(G^0)$ where G^0 is the connected component of the identity in G.

Now a linear algebraic group G is an affine subset of $M_n(K)$ so it is defined by an ideal $I \lhd R$ of the polynomial ring $K[X_{11}, \ldots, X_{nn}]$. In this setting there is a concrete description of $L(G)$. It is a Lie subalgebra of the Lie algebra $M_n(K)$ with the Lie bracket

$$[A, B] = AB - BA.$$

As a vector space, $L(G)$ is the tangent space at the identity element $e \in G$. In our situation this is defined as follows.

For a polynomial $P \in R = K[(x_{ij})]$ and $g = (g_{ij}) \in G \leq M_n(K)$, let ∂P_g be the linear functional on n^2 variables X_{ij} defined as follows

$$\partial P_g : M_n(K) \to K, \quad \partial P_g((X_{ij})_{i,j}) := \sum_{i,j} \left(\frac{\partial P}{\partial x_{ij}}(g_{ij}) \cdot X_{ij} \right).$$

Then $L(G)$ is the subspace of $M_n(K)$ of common solutions to the equations

$$\partial P_e = 0, \quad \forall P \in I,$$

where $e = \mathrm{Id}_n$ is the identity matrix in $G \leq \mathrm{GL}_n(K)$.

In fact, we do not need to check infinitely many equations. By the Hilbert basis theorem, the ideal I is finitely generated, say by polynomials $P_1, \ldots, P_t$. Then $L(G)$ is the common zeros of the linear functionals $\partial(P_i)_e = 0$ $(i = 1, \ldots, t)$.

Connection with Lie algebras of locally compact topological groups

Let $G \leq GL_n$ be a linear algebraic group and suppose that k is a complete field, for example $\mathbb{C}, \mathbb{R}$ or the field of p-adic numbers $\mathbb{Q}_p$ (see example 3.3 below). We have another topology on $\mathrm{GL}_n(k)$ by considering it as a subset of the topological space $M_n(k) = k^{n^2}$. In this way the group G_k of k-rational points is a non-discrete locally compact topological group, by virtue of being a closed subgroup of $\mathrm{GL}_n(k)$. In fact, G_k is a complex or real Lie group when $k = \mathbb{C}$ or $\mathbb{R}$, and is a p-adic analytic group when $k = \mathbb{Q}_p$. In this section we shall use the term *analytic group* to refer to either a Lie group or a p-adic analytic group.

There is a standard way to associate a Lie algebra $L(\mathcal{G})$ to any (complex or real) Lie group $\mathcal{G}$ and as explained in Chapter I such a Lie algebra exists for any p-adic analytic group. One uniform way to define them is as the tangent space at the identity of $\mathcal{G}$. The Lie bracket is the differential of the commutator map in $\mathcal{G}$.

The following Proposition is thus almost self evident:

Proposition 2.11. *If the field k is one from $\mathbb{C}, \mathbb{R}$ or $\mathbb{Q}_p$, then the k-rational points of $L(G)$ (namely, $L(G)_k = L(G) \cap M_n(k)$) coincide with the Lie algebra of the analytic group $\mathcal{G} = G_k$.*

For later use we record another basic result. First observe that when we have an analytic group $\mathcal{G} \leq \mathrm{GL}_n(k)$ with a faithful linear representation in $\mathrm{GL}_n(k)$, then we can also consider the Zariski topology on $\mathcal{G}$ as a subset of GL_n.

Proposition 2.12. *Suppose that the group H is a Zariski dense subgroup of the analytic group $G_k \leq \mathrm{GL}_n(k)$ for G and k as above. Let Ad be the adjoint action of G on its Lie algebra $L(G)$. Then $\mathrm{Ad}(H)_k$ and $\mathrm{Ad}(G)_k$ have the same span in the vector space $\mathrm{End}_k L(G)_k$ over k.*

Moreover, when H is an analytic Zariski-dense in G, the Lie algebra $L(\mathcal{H})$ of $\mathcal{H}$ is an ideal of the Lie algebra $L(G)_k$ of G_k.

Proof. The adjoint action of G on $L(G)$ is given by a set of polynomials (it coincides with the conjugation action of G on $L(G)$ as a subset of $M_n(K)$), and so the map $\mathrm{Ad} : G \to \mathrm{End}_K(L(G))$ is a morphism of algebraic varieties, hence a representation of G as an algebraic group. Since H is Zariski-dense in G_k, it follows that $\mathrm{Ad}(H)$ is Zariski-dense in $\mathrm{Ad}(G_k)$ as subsets of $\mathrm{End}_k L(G)_k$. Since a vector subspace of $\mathrm{End}_k L(G)_k$ is Zariski closed, the first part of the proposition follows immediately.

By a standard result of Lie theory, $L(\mathcal{H})$ is a Lie subalgebra of $L(G)_k$, which is $\mathrm{Ad}(\mathcal{H})$-invariant. Now the stabiliser $\mathrm{Stab}(L(\mathcal{H}))$ of $L(\mathcal{H})$ in $\mathrm{End}_k L(G)_k$ is a subspace of $\mathrm{End}_k L(G)_k$. Since this stabiliser contains $\mathrm{Ad}(\mathcal{H})$, it should also contain $\mathrm{Ad}(G_k)$ by the first part of the proposition. Therefore, $L(\mathcal{H})$ is $\mathrm{Ad}(G_k)$ invariant and so it is an ideal of the Lie algebra $L(G)_k$ of G_k. $\qquad\square$

Note: The Lie algebra is a local tool, it was only defined from a neighbourhood of the identity of an analytic group $\mathcal{G}$. So it is the same for any open subgroup of $\mathcal{G}$. In particular, any subgroup of finite index in $G_{\mathbb{Z}_p}$ is a compact open subgroup of $G_{\mathbb{Q}_p}$ and hence all of these groups share the same Lie algebra as analytic groups. In fact, this property characterises the open subgroups of analytic groups:

Proposition 2.13. *Suppose that the analytic group $\mathcal{H}$ is a closed subgroup of the analytic group $\mathcal{G}$. Then $\mathcal{H}$ is an open subgroup of $\mathcal{G}$ if an only if $\mathcal{H}$ and $\mathcal{G}$ have the same Lie algebra. In particular, when $\mathcal{G}$ is compact this happens if and only if $\mathcal{H}$ has finite index in $\mathcal{G}$.*

As in the theory of Lie groups, the Lie algebra is a very useful tool in the study of algebraic groups. This is best seen in the classification of the simple algebraic groups in next section, but we can already give a non-trivial application.

Proposition 2.14. *A connected linear algebraic group G of dimension less than 3 is soluble. If $\dim G = 1$, then G is abelian.*

Indeed, $L = L(G)$ is a Lie algebra of dimension at most 2 as a vector space over K and it is easy to see that L must be soluble. If $\dim L = 1$, then L is abelian and then so is G.

Note that even at this small dimension we see that two connected groups (for example $\mathbb{G}_+$ and $\mathbb{G}_\times$) may have the same Lie algebra and still be non-isomorphic. However, the simply connected, semisimple groups are indeed uniquely determined by their Lie algebras, as we shall see in the following section.

2.3 Semisimple algebraic groups: the classification of simply connected algebraic groups over K

Definition 2.15. A connected algebraic group is called *semisimple* if it has no non-trivial closed connected soluble normal subgroups.

In general, an algebraic group G has a unique maximal connected soluble normal subgroup. This is called the (soluble) *radical* and denoted $\mathrm{Rad}(G)$. The group $G/\mathrm{Rad}(G)$ is then semisimple.

Definition 2.16. A connected algebraic group is *simple* if it is non-abelian and has no proper non-trivial connected normal subgroups.

This implies that every closed proper normal subgroup of G is central and finite (*Exercise*: prove this!).

Theorem 2.17. *A semisimple group G is a central product*

$$G \cong S_1 \circ S_2 \circ \cdots \circ S_l$$

of simple algebraic groups S_i. The factors in this product are unique up to reordering.

Recall that a central product $S_1 \circ S_2 \circ \cdots \circ S_l$ is a quotient P/N of the direct product $P = S_1 \times \cdots \times S_l$ by a central subgroup N intersecting each S_i trivially.

So in order to understand semisimple algebraic groups it is sufficient to understand simple algebraic groups and their central extensions.

Analogous definitions apply relative to any field of definition k. A connected non-abelian algebraic group defined over k is *k-simple* (respectively *k-semisimple*) if it has no non-trivial closed connected proper (respectively soluble) normal subgroups defined over k. Again a k-semisimple group is k-isomorphic to a central product of k-simple groups, which are unique up to reordering.

When we speak of simple/semisimple groups without indicating the field the understanding is that it is K. In this case, G is called *absolutely* simple (respectively semisimple). *Warning*: a k-simple group need not be absolutely simple (though it is semisimple); see Example 2.20 below.

The classification of absolutely simple algebraic groups mirrors entirely the classification of the finite-dimensional simple Lie algebras over K. Indeed, a simple group G has finite centre and $G/Z(G)$ embeds via Ad as a group of automorphisms of its Lie algebra $L = L(G)$. So once the algebra $L(G)$ is known, the group G is determined up to an *isogeny* as a closed subgroup of $\mathrm{Aut}(L)$. More precisely, we have the following classification theorem:

Theorem 2.18 (Chevalley). *For each Lie type $\mathfrak{X}$ from the list*

$$A_n \ (n \geq 1), \ B_n \ (n \geq 2), \ C_n \ (n \geq 3), \ D_n \ (n \geq 4), \ G_2, \ F_4, \ E_6, \ E_7, \ E_8$$

there are two distinguished simple groups of type $\mathfrak{X}$: the so-called simply connected *group G_{sc} and the* adjoint *group $G_{ad} = G_{sc}/Z(G_{sc})$. Every simple group of type $\mathfrak{X}$ is an image of G_{sc} modulo a finite central subgroup T. Such a quotient map $\pi : G \to G/T$ is called a (central) isogeny; all the groups of the same type $\mathfrak{X}$ form one isogeny class.*

Every simple algebraic group belongs to exactly one of the isogeny classes described above.

The proof of the uniqueness of the isogeny classes can be found in [7, Ch. XI] (see Theorem' in 32.1 there). Their *existence* is discussed briefly in [7, 33.6] and the construction of the groups of adjoint types is given in [4].

Examples of simply connected groups are $SL_n(K)$ of type A_{n-1} and $Sp_{2n}(K)$ (type C_n). The group $SO_n(K)$ is simple of type $B_{(n-1)/2}$ or $D_{n/2}$ (depending on whether n is even or odd) but is not simply connected: its universal cover (i.e. the simply connected group in its isogeny class) is $Spin_n(K)$, the so-called spinor group.

We extend the definition of 'simply connected' to the semisimple groups:

Definition 2.19. A semisimple group is simply connected if it is a direct product of simply connected, simple groups.

From Theorem 2.18 it now follows that each semisimple group is the image by a central isogeny of a unique simply connected, semisimple group.

In general, the k-simple algebraic groups are not so easy to describe. In the first place, the radical of such a group is defined over k and so it must be trivial. Therefore, a k-simple (even a k-semisimple) group is also absolutely semisimple.

The next example gives a $\mathbb{Q}$-simple group which is not absolutely simple.

Example 2.20. Let $H = \mathcal{R}_{\mathbb{Q}(i)/\mathbb{Q}}SL_2$ be the restriction of scalars of SL_2 (defined over $\mathbb{Q}$) from $\mathbb{Q}(i)$ to $\mathbb{Q}$. Then H is $\mathbb{Q}$-simple.

Indeed, by Exercise 8.6 on page 94 we see that H is $\mathbb{Q}(i)$-isomorphic to $SL_2 \times SL_2$ via an isomorphism ρ, say. Composing ρ with complex conjugation τ has the effect of swapping the two factors SL_2. It follows that none of these two factors is $\mathbb{Q}$-defined as a subgroup of H. Now if H had a proper $\mathbb{Q}$-defined normal subgroup L, then L must coincide with one of the two factors SL_2 but they are not defined over $\mathbb{Q}$. Contradiction, hence H is $\mathbb{Q}$-simple.

Suppose now that G is a connected, simply connected, k-simple group. This means that over K our group G is isomorphic to a direct product $\prod_i H_i$ of simply connected, K-simple groups H_i. It happens that each of H_i is defined over some finite Galois extension k_1 of k and we have that G is k-isomorphic to the restriction of scalars $\mathcal{R}_{k_1/k}H$ where $H = H_1$, say.

The group H is K-simple, so over K it is isomorphic to one of the (simply connected) groups listed in Theorem 2.18, but we need to classify such groups up to k_1-isomorphism.

For example, the group

$$SO_2 = \left\{ \begin{pmatrix} a & b \\ -b & a \end{pmatrix} \;\middle|\; a^2 + b^2 = 1 \right\}$$

is isomorphic to the multiplicative group $\mathbb{C}_\times$ over K, but this isomorphism is not defined over the real subfield $\mathbb{R}$. Another important family of examples arises from quadratic forms or division algebras as follows:

Example 2.21. Let $V = \mathbb{R}^{2n}$ be the $2n$-dimensional vector space over the reals, let $k \in \{0, 1, \ldots, n\}$ and let $f : V \to \mathbb{R}$ be quadratic form $f(x_1, \ldots, x_{2n}) = \sum_{i=1}^k x_i^2 - \sum_{i=k+1}^{2n} x_i^2$. Let $SO(f) = SO_{2n}(f)$ be the algebraic group of matrices of determinant 1 which preserve f, i.e.

$$SO(f) = \{x \in GL_{2n} \mid \det x = 1 \text{ and } f(xv) = f(v) \; \forall v \in V\}.$$

Then $\mathrm{SO}(f)$ is defined over $\mathbb{R}$ (in fact, it is defined over $\mathbb{Z}$) and is isomorphic to SO_{2n} over $\mathbb{C}$, but for each $k = 0, 1, \ldots, n$ the groups $\mathrm{SO}(f)$ are pairwise non-isomorphic over $\mathbb{R}$. The Witt index of the form f is k (this is the maximal dimension of a subspace W of V such that $f|_W = 0$). From this fact it is easy to see that the maximal rank of a $\mathbb{R}$-split torus of $\mathrm{SO}(f)$ is k. (See Definition 2.24 of a split torus in the next section.) So the groups $\mathrm{SO}(f)$ are pairwise non-isomorphic over $\mathbb{R}$ for $k = 0, 1, \ldots, n$.

Example 2.22. Let H be the 4-dimensional quaternion algebra over $\mathbb{R}$. We can consider H as a subalgebra of $M_4(\mathbb{R})$ via the regular representation. For an integer $n \in \mathbb{N}$, consider the group $\mathrm{GL}_n(H)$ of invertible n by n matrices with entries in H. We can consider $\mathrm{GL}_n(H)$ as a subset of $M_{4n}(\mathbb{R})$ and define $\mathrm{SL}_n(H)$ to be the subgroup of matrices with determinant 1, i.e. $\mathrm{SL}_n(H) = \mathrm{GL}_n(H) \cap \mathrm{SL}_{4n}(\mathbb{R})$. Then $\mathrm{SL}_n(H)$ is a closed subgroup of $\mathrm{SL}_{4n}(\mathbb{R})$ and it is easy to see that it is $\mathbb{C}$-isomorphic to SL_{2n} (because $H \otimes_{\mathbb{R}} \mathbb{C} \simeq M_2(\mathbb{C})$). However, $\mathrm{SL}_n(H)$ is not $\mathbb{R}$-isomorphic to SL_{2n} because the maximal rank of a $\mathbb{R}$-split torus in it is $n - 1$.

The k_1-isomorphism classes of groups which are K-isomorphic to H are called the k_1-forms of H. These are classified by the non-commutative 1-cohomology set $H^1(\mathrm{Gal}(K/k_1), \mathrm{Aut} H_{ad})$. For example, the unitary group SU_n is isomorphic to SL_n over $K = \mathbb{C}$ but not over $\mathbb{R}$, and together with examples 2.22 these are different real forms of SL_n. Similarly, the group G in Exercise 8.7 on page 94 is a $\mathbb{Q}$-form of SL_2, while the groups from example 2.21 are real forms of SO_{2n}. For more details about the classification the $\mathbb{Q}$-forms of classical groups, we refer to [16, Ch. 2]. There is an alternative classification by Borel and Tits based on the concept of *relative root systems*, for that see [19].

2.4   Reductive groups

A class of groups which is more general than semisimple groups but which shares some of their nicer properties is the *reductive groups*. For example, $\mathrm{GL}_n(K)$ is not semisimple but still a very important group which we would like to include in our theory.

Definition 2.23. An element g of a linear algebraic group $G \leq \mathrm{GL}_n(K)$ is called *unipotent* (respectively *semisimple*) if g is unipotent (respectively diagonalisable) as a matrix in $\mathrm{GL}_n(K)$. This definition is independent of the choice of the linear representation of G. The group G is *unipotent* if it consists of unipotent elements.

For example, $\mathbb{G}_+$ is unipotent.

Now it can be shown that an algebraic group G has a unique maximal normal unipotent subgroup. This is the *unipotent radical* of G and is denoted $R_u(G)$. The group G is called *reductive* if $R_u(G) = 1$. One obvious example of a reductive group is a torus.

Definition 2.24. An algebraic group T is a *torus* if it is isomorphic to a direct product $\mathbb{G}_\times^m$. The *rank* of T is the number m of direct factors $\mathbb{G}_\times$. The torus T is called *k-split* if there is a k-defined isomorphism $T \to \mathbb{G}_\times^m$.

The group of diagonal matrices in $\mathrm{GL}_n(K)$ is a torus of rank n.

Theorem 2.25. *A connected reductive group G is a product $G = TS$ of a torus T and a semisimple subgroup S such that $[T, S] = 1$ and $T \cap S$ is finite. The subgroups T and S are unique.*

For example, $\mathrm{GL}_n(K)$ is reductive and equal to $Z \cdot \mathrm{SL}_n(K)$ where the torus $Z \cong \mathbb{G}_\times$ is the group of scalar matrices.

2.5   Chevalley groups

Definition 2.26. Let G be an algebraic group defined over a field k. Then G is called *k-split* if it has a maximal torus which is k-split.

There is a simply connected, unique simple and $\mathbb{Q}$-split algebraic group of any given Lie type $\mathfrak{X}$ and this is called the *Chevalley group of type $\mathfrak{X}$*. There is a simple conceptual way to define the adjoint group $\overline{G} = G/Z(G)$, as described for example in [4, Ch. 1]. As we have seen, $\overline{G}$ acts faithfully on the Lie algebra $L = L(G)$ of G and so can be identified with a subgroup of $\mathrm{Aut}(L)$. In fact, $\overline{G}$ is defined as the subgroup of $\mathrm{Aut}(L)$ generated by the elements

$$\exp(\mathrm{ad}(x)) = 1 + \mathrm{ad}(x) + \frac{\mathrm{ad}(x)^2}{2!} + \frac{\mathrm{ad}(x)^3}{3!} + \cdots,$$

where x is an element of a root subgroup of L. Note that for such x the linear transformation, $\mathrm{ad}(x) : L \to L$ is nilpotent and so the above series is finite.

Moreover, as described in [4] one can find a Lie subring S of L which is a finitely generated torsion free $\mathbb{Z}$-lattice of L and such that $\exp(\mathrm{ad}(x))$ stabilises S for each x as above. Hence, $\overline{G}$ is in fact defined over $\mathbb{Z}$ and one sees that the same is true for the universal cover G. Therefore, its R-rational points G_R are defined for any ring R. In particular, $G_\mathbb{F}$ is defined for any finite field $\mathbb{F}$. As we shall see in Section 6.1, this is the construction of the untwisted finite simple groups of Lie type.

3   Arithmetic groups and the congruence topology

In this section and below, k will denote a number field (a finite extension of $\mathbb{Q}$) and $\mathcal{O}$ its ring of integers. By convention, *prime ideals* of $\mathcal{O}$ are assumed non-zero. We begin by recalling some information about the ring $\mathcal{O}$.

3.1   Rings of algebraic integers in number fields

$\mathcal{O}$ is the collection of all elements $x \in k$ satisfying a polynomial equation

$$x^n + a_1 x^{n-1} + \cdots + a_{n-1} x + a_n = 0$$

with leading coefficient 1 and each $a_i \in \mathbb{Z}$. This is in fact a subring of k. As an additive group it is isomorphic to $\mathbb{Z}^d$, the free abelian group of rank d, where $d = (k : \mathbb{Q})$.

The ring $\mathcal{O}$ has Krull dimension 1, i.e. every prime ideal is maximal. Moreover, every non-zero ideal has finite index in $\mathcal{O}$. Each non-zero ideal I can be factorised

$$I = \mathfrak{p}_1^{e_1} \cdot \mathfrak{p}_2^{e_2} \cdots \mathfrak{p}_m^{e_m}$$

as a product of prime ideals $\mathfrak{p}_i$ and this factorisation is unique up to reordering of the factors. The Chinese remainder theorem says that then

$$\mathcal{O}/I \cong \mathcal{O}/\mathfrak{p}_1^{e_1} \oplus \mathcal{O}/\mathfrak{p}_2^{e_2} \oplus \cdots \oplus \mathcal{O}/\mathfrak{p}_m^{e_m}.$$

Each prime ideal $\mathfrak{p}$ divides (i.e. contains) a unique rational prime $p \in \mathbb{N}$, and then $\mathfrak{p} \cap \mathbb{Z} = p\mathbb{Z}$. The quotient $\mathcal{O}/\mathfrak{p}$ is a finite field of characteristic p.

If $p\mathcal{O} = \prod_{i=1}^{g} \mathfrak{p}_i^{e_i}$ is the factorisation of the principal ideal $(p) = p\mathcal{O}$, then

$$d = (k : \mathbb{Q}) = \sum_{i=1}^{g} e_i n_i, \quad \text{where } |\mathcal{O}/\mathfrak{p}_i| = p^{n_i}. \tag{3.1}$$

If k is a Galois extension of $\mathbb{Q}$, then $e_1 = \cdots = e_g$ and $n_1 = \cdots = n_g$. Also $e_i \neq 1$ for at most finitely many rational primes p (the so-called *ramified* ones). The **Chebotarev's density theorem** (see [16, Ch. 1]) implies that for a positive proportion of all rational primes p we have $g = (k : \mathbb{Q})$ and $n_1 = \cdots n_g = 1$, i.e. the ideal (p) splits in k. More precisely

$$\lim_{x \to \infty} \frac{|\{p \leq x \mid p \text{ a prime which splits in } k\}|}{|\{p \leq x \mid p \text{ prime }\}|} = \frac{1}{|\mathrm{Gal}(k/\mathbb{Q})|} = \frac{1}{d}.$$

Here $\mathrm{Gal}(k/\mathbb{Q})$ is the Galois group of k over $\mathbb{Q}$.

Let S be a finite set of prime ideals. An element $a \in k$ is said to be *S-integral* if $Ja \subseteq \mathcal{O}$ where J is some product of prime ideals in S. The set of all S-integral elements forms a subring $\mathcal{O}_S$ of k, containing $\mathcal{O}$, called the ring of *S-integers* of k. Of course, when S is empty $\mathcal{O}_S = \mathcal{O}$.

3.2   The congruence topology on $\mathrm{GL}_n(k)$ and $\mathrm{GL}_n(\mathcal{O})$

The *congruence topology* on k has as a base of open neighbourhoods of 0 the set of all non-zero ideals of $\mathcal{O}$. The congruence topology on $M_n(k) = k^{n^2}$ is then the product topology, and the congruence topology on $\mathrm{GL}_n(k)$ (and on any closed subgroup) is the one induced by that on $M_n(k)$. This means that a base of neighbourhoods of 1 is the set of subgroups $\mathrm{GL}_d(k) \cap (1_n + M_n(I))$ with I a non-zero ideal of $\mathcal{O}$.

More generally, for any set S of prime ideals, we define the *S-congruence topology* by taking only ideals that are products of prime ideals not in S; equivalently, we can take as a neighbourhood basis the set of all non-zero ideals of $\mathcal{O}_S$.

It is easy to see that the congruence topology on k and hence on $M_n(k)$ is Hausdorff: if $x \neq y$ are two elements of k, then there is an ideal I of $\mathcal{O}$ such that $x - y \notin I$ and hence $(x + I) \cap (y + I) = \emptyset$. In fact, the congruence topology on $M_n(k)$ is finer than the Zariski topology as the following proposition demonstrates.

Proposition 3.1. *Let W be a k-defined Zariski closed set of $M_n(K) = K^{n^2}$ defined by an ideal T of polynomials in its n^2 coordinates. Then W_k is closed in the congruence topology of $M_n(k)$.*

Proof. Let $\mathbf{x} \in M_n(k)$ be an element of the congruence closure of W_k. So for any ideal I of $\mathcal{O}$ we have an element $\mathbf{y} \in W_k$ such that $\mathbf{x} \equiv \mathbf{y}$ mod I. Now let p be a polynomial from T with coefficients in k. We may assume that up to a scalar multiple, p has coefficients from $\mathcal{O}$. But then $p(\mathbf{x}) \equiv p(\mathbf{y}) = 0$ mod I, so $p(\mathbf{x}) \in I$ for any ideal I of $\mathcal{O}$. This is possible only if $p(\mathbf{x}) = 0$. Since this holds for all polynomials p with coefficients in k, and since W is defined over k, we deduce that $\mathbf{x} \in W$. $\qquad\square$

Example 3.2. Let $k = \mathbb{Q}$. Then any finite union or intersection of sets of the form

$$\{a + m\mathbb{Z}\} \times \{b + n\mathbb{Z}\} \subset \mathbb{Q}^2 \quad a, b, m, n \in \mathbb{Z}$$

is an open set in the congruence topology of $\mathbb{Q}^2$ but none of them is Zariski open.

It is thus clear that the congruence topology on $M_n(k)$ has many more closed sets than the Zariski topology. So a Zariski dense subset of matrices may be rather 'sparse' in the congruence topology. From this point of view it is indeed surprising that in the case of simple algebraic groups the property of being Zariski dense has a rather strong implication for the congruence closure of a subgroup. This is the main content of Theorem 5.1 below.

Valuations of k

For any prime ideal $\mathfrak{p}$ of $\mathcal{O}$, the $\mathfrak{p}$-adic topology is defined in the same way as the congruence topology except that the ideals are only allowed to be positive powers of $\mathfrak{p}$. The completion of k with respect to this topology is denoted $k_{\mathfrak{p}}$ and the closure of $\mathcal{O}$ in $k_{\mathfrak{p}}$ is denoted $\mathcal{O}_{\mathfrak{p}}$.

The valuation $v_{\mathfrak{p}}$ on $k_{\mathfrak{p}}$ is defined by $v_{\mathfrak{p}}(a) = t$ where $t \in \mathbb{Z}$ is the largest integer such that $\mathfrak{p}^{-t}a \subseteq \mathcal{O}_{\mathfrak{p}}$ (if $a \neq 0$, one sets $v_{\mathfrak{p}}(0) = \infty$). Thus $\mathcal{O}_{\mathfrak{p}}$ is the *valuation ring*, consisting of all elements of k having valuation ≥ 0; this implies that $\mathcal{O}_{\mathfrak{p}}$ is a local ring, having $\mathfrak{p}\mathcal{O}_{\mathfrak{p}}$ as its unique maximal ideal. (One often associates to such a valuation $v_{\mathfrak{p}}$ the corresponding *absolute value*: $|a|_{\mathfrak{p}} = q^{-v_{\mathfrak{p}}(a)}$ where $q = |\mathcal{O}/\mathfrak{p}|$, which is multiplicative.)

Example 3.3 (The p-adic numbers). Take $k = \mathbb{Q}$ with a ring of integers $\mathbb{Z}$. Let p be a prime. The *p-adic* valuation $v_p(x)$ is the usual one where $v_p(x) = t$ is the largest integer such that $x = p^t a/b$ with integers a and b coprime to p. A base for neighbourhoods of 0 in the p-adic topology on $\mathbb{Q}$ is the family of subgroups $\left\{\frac{p^l a}{b} \mid a, b \in \mathbb{Z}, (p, b) = 1\right\}$, $l = 1, 2, \ldots$. The completion of $\mathbb{Q}$ with respect to this topology is the field $\mathbb{Q}_p$ of *p-adic numbers*. Inside $\mathbb{Q}_p$ we have the closure $\mathbb{Z}_p$ of $\mathbb{Z}$, which is the ring of *p-adic integers*. We can view $\mathbb{Z}_p$ as a ring of infinite power series in p

$$a_0 + a_1 p + \cdots + a_k p^k + \cdots, \quad a_i \in \{0, 1, \ldots, p-1\}$$

with the obvious addition and multiplication. The finite such sums comprise the subring $\mathbb{Z}$. The unique maximal ideal is $p\mathbb{Z}_p$ and every element $x \in \mathbb{Q}_p$ can be written uniquely as $x = p^t y$ for some $y \in \mathbb{Z}_p \smallsetminus p\mathbb{Z}_p$ and $t \in \mathbb{Z}$.

In general, if $\mathfrak{p}_i$ is a prime ideal of $\mathcal{O}$ dividing p, then $k_{\mathfrak{p}_i}$ is a vector space over $\mathbb{Q}_p$ of dimension $e_i n_i$, the number in (3.1) above.

The valuations $v_{\mathfrak{p}}$ for all prime ideals $\mathfrak{p}$ of $\mathcal{O}$ comprise the *non-archimedean* valuations of k. Now, suppose that the number field k has s embeddings $v_i : k \to \mathbb{R}$ $(i = 1, \ldots, s)$ and $2t$ non-real embeddings $v_j, \bar{v}_j : k \to \mathbb{C}$ $(j = s+1, \ldots, s+t)$. Composing these with the ordinary real or complex absolute value gives the set V_∞ of $s + t$ *archimedean* absolute values on k. For $v \in V_\infty$, we put $k_v = \mathbb{R}$ or $k_v = \mathbb{C}$ according to whether the corresponding embedding of k is real or non-real.

The ring of S-integers has a more natural definition in terms of valuations

$$\mathcal{O}_S = k \cap \bigcap_{\mathfrak{p} \notin S} \mathcal{O}_{\mathfrak{p}}.$$

A word of warning: the notation $\mathcal{O}_S$ can be a bit confusing: if $S = \{\mathfrak{q}\}$ consists of a single prime, then $\mathcal{O}_{\{\mathfrak{q}\}}$ is the ring of $\{\mathfrak{q}\}$-integers, while $\mathcal{O}_{\mathfrak{q}}$ is the completion of $\mathcal{O}$ at $\mathfrak{q}$. For example, $\mathbb{Z}_{\{p\}} = \mathbb{Z}[1/p] \subset \mathbb{Q}$, while $\mathbb{Z}_p$ is the ring of p-adic integers.

3.3   Arithmetic groups

Suppose we are given a linear algebraic group G defined over k with a faithful representation $G \hookrightarrow \mathrm{GL}_n(K)$, also defined over k.

Definition 3.4. A subgroup Γ of G_k is called *arithmetic* if it is commensurable with the group of $\mathcal{O}$-integral points $G_\mathcal{O}$ (in other words, $\Gamma \cap G_\mathcal{O}$ has finite index in both Γ and $G_\mathcal{O}$).

It turns out that this definition is independent of the choice of k-defined linear representation of G.

More generally, we can define the *S-arithmetic subgroups* of G_k as those commensurable with $G_{\mathcal{O}_S}$. When the set S has not been specified we shall always assume that it is empty.

The simplest examples of arithmetic groups are $(\mathcal{O}, +)$ and $(\mathcal{O}^*, \times)$, the additive and multiplicative groups of the ring of integers of k. We thus see that the study of arithmetic groups is a generalisation of classical algebraic number theory.

One of the most general results about arithmetic groups is the following:

Theorem 3.5 ([16, Ch. 4]). *Let Γ be an arithmetic subgroup of a k-defined linear algebraic group G as above. Then Γ is finitely presented and has only finitely many conjugacy classes of finite subgroups.*

For S-arithmetic groups, the above statement is still true, provided that G is reductive.

Now an $(S\text{-})$arithmetic group Γ has its own $(S\text{-})$congruence topology induced from the $(S\text{-})$congruence topology of $\mathrm{GL}_n(k)$. We call a subgroup $\Delta \leq \Gamma$ an $(S\text{-})$congruence subgroup if it is open in this topology, i.e. if Δ contains a *principal congruence subgroup* $\Gamma \cap (1_n + M_n(I))$ for some non-zero ideal I of (coprime to S). The *congruence images* Γ/N of Γ are those with the kernel a congruence subgroup $N \lhd \Gamma$.

Clearly, a congruence subgroup of Γ has finite index, but the converse is not true in general. When it does hold, i.e. if every subgroup of finite index is a congruence subgroup, Γ is said to have *the congruence subgroup property* (CSP).

There is a neat way to state CSP in term of profinite groups. If $\mathfrak{X}$ is an intersection-closed family of normal subgroups of finite index in Γ, one defines the $\mathfrak{X}$-*completion* of Γ to be the inverse limit

$$\widehat{\Gamma}_{\mathfrak{X}} = \varprojlim_{N \in \mathfrak{X}} \Gamma/N$$
$$= \{(\gamma_N)_{N \in \mathfrak{X}} \mid p_{NM}(\gamma_N) = \gamma_M \; \forall N \leq M \in \mathfrak{X}\} \leq \prod_{N \in \mathfrak{X}} \Gamma/N,$$

where $p_{NM} : \Gamma/N \to \Gamma/M$ denotes the natural quotient map for each $N \leq M$. (With the topology induced from the product topology on the Cartesian product, $\widehat{\Gamma}_{\mathfrak{X}}$ becomes a compact topological group, a *profinite group*).

A natural example of inverse limits is the valuation ring $\mathcal{O}_{\mathfrak{p}}$: for a prime ideal $\mathfrak{p}$ of $\mathcal{O}$, the inverse limit

$$\varprojlim_{n \in \mathbb{N}} \mathcal{O}/\mathfrak{p}^n \mathcal{O}$$

is isomorphic as a ring to the completion $\mathcal{O}_{\mathfrak{p}}$ of $\mathcal{O}$ with respect to the $\mathfrak{p}$-adic topology defined by the powers of the ideal $\mathfrak{p}$. This also shows that $\mathcal{O}_{\mathfrak{p}}/\mathfrak{p}^n \mathcal{O}_{\mathfrak{p}}$ is isomorphic to $\mathcal{O}/\mathfrak{p}^n \mathcal{O}$.

We are interested in two special choices for $\mathfrak{X}$. When $\mathfrak{X}$ consists of *all* normal subgroups of finite index, $\widehat{\Gamma}_{\mathfrak{X}} = \widehat{\Gamma}$ is the *profinite completion* of Γ. When $\mathfrak{X}$ consists of all the normal *congruence* subgroups, $\widehat{\Gamma}_{\mathfrak{X}} = \widetilde{\Gamma}$ is the *congruence completion* of Γ. There is an obvious natural projection $\pi : \widehat{\Gamma} \to \widetilde{\Gamma}$, which is clearly surjective.

Now we can reformulate the congruence subgroup property as saying that the map π is bijective. For many purposes, the following generalisation of CSP is more relevant: the arithmetic group Γ is said to have the *generalised congruence subgroup property* (GCSP for short) if the kernel of $\pi : \widehat{\Gamma} \to \widetilde{\Gamma}$ is finite. Group theoretically, this says that any subgroup of finite index in Γ is commensurable 'with bounded index' with a congruence subgroup. There is a famous conjecture by Serre which characterises the S-arithmetic groups (in semisimple algebraic groups) with GCSP as those having S-rank at least 2: see Section 4.1.

4   The strong approximation theorem

The congruence images of the S-arithmetic group $\Gamma = G_{\mathcal{O}_S}$ are easier to understand when G has the *strong approximation property*. In order to explain this, we need several more definitions.

Recall that $k_{\mathfrak{p}}$ and $\mathcal{O}_{\mathfrak{p}}$ are the completions of k and $\mathcal{O}$ with respect to the $\mathfrak{p}$-adic topology defined by powers of the prime ideal $\mathfrak{p} \lhd \mathcal{O}$. As usual we set $G_{k_{\mathfrak{p}}} = G \cap M_n(k_{\mathfrak{p}})$ and $G_{\mathcal{O}_{\mathfrak{p}}} = G \cap M_n(\mathcal{O}_{\mathfrak{p}})$. The first of these is a locally compact totally disconnected topological group and the second is a compact subgroup. In fact, $G_{\mathcal{O}_{\mathfrak{p}}}$ is an example of a *p-adic analytic group*. We refer to $G_{k_{\mathfrak{p}}}$ as the completion of G at $\mathfrak{p}$.

Similarly, if v is an archimedean real (respectively complex) absolute value of k associated to an embedding ν_i, then we write G_v for $G_{\mathbb{R}}^{\nu_i}$ (respectively $G_{\mathbb{C}}^{\nu_i}$) where $G^{\nu} \leq \mathrm{GL}_n(\mathbb{C})$ is the group obtained by applying ν to the defining equations of the affine variety G.

The profinite groups $G_{\mathcal{O}_{\mathfrak{p}}}$ are in close relationship with the congruence images of $G_{\mathcal{O}_S}$.

Recall that the algebraic group G is k-defined. It is easy to see that for almost all prime ideals $\mathfrak{p}$, the coefficients of the equations defining G in GL_n are not divisible by $\mathfrak{p}$. Therefore, we can consider these equations modulo $\mathfrak{p}^n$ for any $n \in \mathbb{N}$. Denote the set of their solutions in $\mathcal{O}/\mathfrak{p}^n$ by $G_{\mathcal{O}/\mathfrak{p}^n}$: this is a finite subgroup of $\mathrm{GL}_n(\mathcal{O}/\mathfrak{p}^n)$ and is called the *reduction* of G modulo $\mathfrak{p}^n$. See [16, pp. 142–146] for more details about reductions of affine algebraic varieties and groups.

Now consider the quotient mapping

$$\mathcal{O}_{\mathfrak{p}} \to \mathcal{O}_{\mathfrak{p}}/\mathfrak{p}^n\mathcal{O}_{\mathfrak{p}} \simeq \mathcal{O}/\mathfrak{p}^n\mathcal{O}.$$

This induces a homomorphism

$$\pi_{\mathfrak{p}^n} : G_{\mathcal{O}_{\mathfrak{p}}} \to G_{\mathcal{O}/\mathfrak{p}^n\mathcal{O}}.$$

The following is the content of Proposition 3.20 of [16].

Proposition 4.1. *The maps $\pi_{\mathfrak{p}^n}$ are surjective for all but finitely many primes* $\mathfrak{p}$ *(and all integers n).*

We say that G has *good reduction* for such primes p.

Assume from now on that $\mathfrak{p}$ is not in the finite set S. The restriction of $\pi_{\mathfrak{p}^n}$ to its dense subgroup $G_{\mathcal{O}_S} \leq G_{\mathcal{O}_\mathfrak{p}}$ is the homomorphism

$$G_{\mathcal{O}_S} \to G_{\mathcal{O}_S/\mathfrak{p}^n \mathcal{O}_S}$$

obtained by reducing all entries of $\Gamma = G_{\mathcal{O}_S} \leq \mathrm{GL}_n(\mathcal{O}_S)$ modulo $\mathfrak{p}^n$. So the images of $\pi_{\mathfrak{p}^n}$ are all congruence images of Γ. What is not clear at this point is how to combine these to describe the congruence images of Γ at composite ideals. This is the content of the strong approximation theorem below.

Define

$$G_S := \prod_{v \in V_\infty} G_v \times \prod_{p \in S} G_{k_p}.$$

This is a locally compact group and the image of Γ in G_S under the diagonal embedding in each factor is a *lattice* in G_S, i.e. a discrete subgroup of finite co-volume. As a consequence, the arithmetic subgroup $\Gamma = G_{\mathcal{O}_S}$ is infinite if and only if the group G_S is non-compact.

Let

$$G_{\widehat{\mathcal{O}}_S} = \prod_{p \notin S} G_{\mathcal{O}_p}.$$

Again there is an obvious diagonal embedding $i : \Gamma \to G_{\widehat{\mathcal{O}}_S}$ and the congruence topology of Γ coincides with the topology induced in $i(\Gamma)$ as a subgroup of the profinite group $G_{\widehat{\mathcal{O}}_S}$. Hence the congruence completion $\widetilde{\Gamma}$ is isomorphic to the closure $\overline{i(G)}$ of $i(G)$ in $G_{\widehat{\mathcal{O}}_S}$. The strong approximation theorem states that under certain conditions $i(G)$ is *dense* in $G_{\widehat{\mathcal{O}}_S}$, and therefore $\widetilde{\Gamma} \simeq G_{\widehat{\mathcal{O}}_S}$.

Theorem 4.2 (*Strong approximation for arithmetic groups*) ([16, Theorem 7.12]). *Let G be a connected, simply connected, simple algebraic group defined over a number field k and let the groups $\Gamma = G_{\mathcal{O}_S}$, $G_S, G_{\widehat{\mathcal{O}}_S}$ and the embedding $i : \Gamma \to G_{\widehat{\mathcal{O}}_S}$ be as above. Assume that Γ is infinite (which is equivalent to G_S being non-compact). Then $i(\Gamma)$ is dense in $G_{\widehat{\mathcal{O}}_S}$ and hence $\widetilde{\Gamma} \simeq G_{\widehat{\mathcal{O}}_S}$.*

When the conclusion holds, we say that $G_{\mathcal{O}_S}$ has the **strong approximation property**, or that G has the strong approximation property w.r.t. S.

Note: Usually the strong approximation theorem is formulated for the group of k-rational points G_k and says that G_k is dense in the adelic group $G_{\mathcal{A}_S}$; the statement we have given above is equivalent to this (and more transparent for arithmetic groups); see [16, Ch. 7].

More generally, a connected algebraic group G has the strong approximation property if its maximal reductive quotient $H = G/R_u(G)$ is a direct product of simply connected, simple groups, and H_S is non-compact.

The strong approximation theorem can be viewed as a generalisation of the Chinese remainder theorem, which in this setting says that the diagonally embedded image of $\mathbb{Z}$ is dense in $\prod_{p \text{ prime}} \mathbb{Z}_p$. In the general situation, the theorem says that the finite images of the product $G_{\widehat{\mathcal{O}}_S}$ coincide with the congruence images of Γ.

Note: The condition that G be simply connected is indeed necessary (Exercise 8.10).

Set $\mathbb{F}_{q(\mathfrak{p})}$ where $q(\mathfrak{p}) = |\mathcal{O}/\mathfrak{p}|$. Then Theorem 4.2 and Proposition 4.1 give:

Corollary 4.3. *Under the hypotheses of Theorem 4.2, we have* $\pi_{\mathfrak{p}}(\Gamma) = G_{\mathbb{F}_{q(\mathfrak{p})}}$ *for all but finitely many primes* $\mathfrak{p} \notin S$.

In turn the groups $G_{\mathbb{F}_{q(\mathfrak{p})}}$ are easy to describe when G is semisimple, see Proposition 6.1 below.

For the moment we shall note case the relationship between G and H when $G = \mathfrak{R}_{k/\mathbb{Q}}H$ is a restiction of scalars of H.

Proposition 4.4. *Let* k *be a finite Galois extension of* $\mathbb{Q}$ *and* $G = \mathfrak{R}_{k/\mathbb{Q}}H$ *be the restriction of scalars of some* $\mathbb{Q}$-*defined algebraic group* H. *Then for almost all primes* p

$$G_{\mathbb{F}_p} = \prod_{\mathfrak{p}|p} H_{\mathbb{F}_{q(\mathfrak{p})}},$$

where the product on the right is over all prime ideals $\mathfrak{p}$ *of* k *dividing* p.

Indeed, all but finitely many primes p are unramified in k and therefore

$$\mathcal{O}/p\mathcal{O} = \prod_{\mathfrak{p}|p} \mathcal{O}/\mathfrak{p} = \prod_{\mathfrak{p}|p} \mathbb{F}_{q(\mathfrak{p})},$$

where $\mathcal{O}$ is the ring of integers of k. The proposition follows immediately from $G_{\mathbb{Z}} = H_{\mathcal{O}}$ and Proposition 4.1.

4.1   An aside: Serre's conjecture

We now have most of the definitions to state Serre's conjecture.

Definition 4.5. For a valuation v of k, the k_v-rank of the topological group G_v is the largest integer n such that G_v contains the direct product $(k_v^*)^n$. The S-rank of an algebraic group G is

$$\sum_{v \in V_\infty \cup S} k_v\text{-rank of } G_v,$$

where V_∞ is the set of all archimedean valuations of k.

Conjecture 4.6 (J.-P. Serre). *A connected, simply connected, simple algebraic group* G *has the generalised* S-*congruence subgroup property if and only if the* S-*rank of* G *is at least 2.*

For example, the group $\mathrm{SL}_n(\mathbb{Z})$ has CSP if $n > 2$ but not if $n = 2$.

Currently Serre's conjecture is open for some groups of S-rank 1 and also when G is a totally anisotropic form of A_n, see [16, §9.5].

5 Lubotzky's alternative

It will be too much to expect that the strong approximation theorem holds for linear groups in general, indeed it does not hold for algebraic tori. Nevertheless, there is something that can be said when the group is non-soluble.

Theorem 5.1 (Nori [15], Weisfeiler [21]). *Let Δ be a Zariski-dense subgroup of a simply connected, $\mathbb{Q}$-simple linear algebraic group $G \le \mathrm{GL}_n(\mathbb{C})$ and suppose that $\Delta \le G_{\mathbb{Z}_S}$ for some finite set of primes S. Let $i : \Delta \to G_{\widehat{\mathbb{Z}}_S}$ be the diagonal embedding. Then the closure $\overline{i(\Delta)}$ of $i(\Delta)$ in $G_{\widehat{\mathbb{Z}}_S}$ is an open subgroup of $G_{\widehat{\mathbb{Z}}_S}$.*

It follows that for all but finitely many primes p, all the groups $G_{\mathbb{Z}/(p^m \mathbb{Z})}$ appear as congruence images of Δ.

There are several different proofs of this theorem. We shall sketch one of them in Section 7. For the moment, let us assume this result and deduce Theorem 1.1. We restate it here:

Theorem 1. *Let $\Delta \le \mathrm{GL}_n(k)$ be a finitely generated linear group over a field k of characteristic 0. Then one of the following holds:*

(a) *the group Δ is virtually soluble, or*

(b) *there exist a connected, simply connected, $\mathbb{Q}$-simple algebraic group G, a finite set of primes S such that $\Gamma = G_{\mathbb{Z}_S}$ is infinite and a subgroup Δ_1 of finite index in Δ such that every congruence image of Γ appears as a quotient of Δ_1.*

Proof of Theorem 1. Suppose that we have a finitely generated linear group $\Delta \le \mathrm{GL}_n(\mathbb{C})$. Then in fact $\Delta \le \mathrm{GL}_n(J)$ for some finitely generated subring J of $\mathbb{C}$.

Now the Jacobson radical (the intersection of the maximal ideals of J) is trivial and so J is residually a number field. Indeed, if $\mathbf{m}$ is a maximal ideal of J, then $J/\mathbf{m}$ is a finitely generated algebra, which is a field. By Corollary 7.10 in [1] ('The weak Nullstellensatz'), $J/\mathbf{m}$ is a finite extension of $\mathbb{Q}$, i.e. a number field.

Hence Δ is residually in $\mathrm{GL}_n(k_i)$ for some number fields k_i. Suppose that Δ is not virtually soluble. By Lemma 1.3 it follows that there is $i \in I$ such that the image of Δ in $\mathrm{GL}_n(k_i)$ is not virtually soluble. Replacing Δ with this image we may assume that $\Delta \le \mathrm{GL}_n(k)$ for some number field k.

Consider $\mathrm{GL}_n(k)$ as a subgroup of $\mathrm{GL}_{nd}(\mathbb{Q})$ where $d = (k : \mathbb{Q})$. Let $\mathcal{G}$ be the Zariski-closure of Δ in $\mathrm{GL}_{nd}(K)$. This is a $\mathbb{Q}$-defined linear algebraic group and we take its connected component $\mathcal{G}_0$ at the identity.

Let $\Delta_1 = \mathcal{G}_0 \cap \Delta$. This has finite index in Δ and is Zariski-dense in $\mathcal{G}_0$. Since Δ is not virtually soluble, the connected algebraic group $\mathcal{G}_0$ is not soluble. By Exercise 8.12, we see that there exists a connected, $\mathbb{Q}$-simple algebraic group G and a $\mathbb{Q}$-defined epimorphism $f : \mathcal{G}_0 \to G$. Now $f(\Delta_1)$ is dense in G and we may replace Δ by $f(\Delta_1)$ and $\mathcal{G}_0$ by G to reduce the situation to where we have a finitely generated Zariski-dense subgroup $\Delta \leq G_\mathbb{Q}$ of a connected, $\mathbb{Q}$-simple linear algebraic group G. The main difference with the setup of Theorem 5.1 is that G may not be simply connected. However, G is isogenous to its simply connected cover $\widetilde{G}$, i.e. there is a $\mathbb{Q}$-defined surjection $\pi : \widetilde{G} \to G$ where $\ker \pi = Z$ is a finite central subgroup of $\widetilde{G}$.

It is not in general true that $\pi(\widetilde{G}_\mathbb{Q}) = G_\mathbb{Q}$ but at least we have the following:

Proposition 5.2. *The group* $G_\mathbb{Q}/\pi(\widetilde{G}_\mathbb{Q})$ *is abelian of finite exponent dividing* $|Z|$.

Proof. Let A be the Galois group of $K/\mathbb{Q}$, where K is the algebraic closure of $\mathbb{Q}$. Then $\widetilde{G}_\mathbb{Q}$ consists of all those $g \in \widetilde{G}_K$ such that $g^\alpha = g$ for all $\alpha \in A$. On the other hand, $\pi^{-1}(G_\mathbb{Q})$ consists of those $g \in \widetilde{G}_K$ such that $g^\alpha \equiv g$ mod Z for all $\alpha \in A$. Suppose that $g, h \in \pi^{-1}(G_\mathbb{Q})$, thus $g^\alpha \equiv g$ and $h^\alpha \equiv h$ mod Z for all $\alpha \in A$. Now using that Z is central in $\widetilde{G}$, we see that $[g,h]^\alpha = [g^\alpha, h^\alpha] = [g,h]$ and hence that $[g,h] \in \widetilde{G}_\mathbb{Q}$. Let $m = \exp Z$. In the same way we see that if $g^\alpha \equiv g$ mod Z, then $(g^m)^\alpha = g^m$ and therefore $g^m \in \widetilde{G}_\mathbb{Q}$. So $\pi^{-1}(G_\mathbb{Q})/\widetilde{G}_\mathbb{Q}$ is abelian of exponent dividing $|Z|$ and this implies the proposition. $\square$

Now take $\Delta_0 = \Delta \cap \pi(\widetilde{G}_\mathbb{Q})$; this is a subgroup of finite index in Δ because Δ/Δ_0 is a finitely generated abelian group of finite exponent. Let $U_0 = \pi^{-1}(\Delta_0) \cap \widetilde{G}_\mathbb{Q}$, then $U_0/(U_0 \cap Z) \simeq \Delta_0$; U_0 is a finitely generated linear group and thus it is residually finite. So we can find a subgroup U of finite index in U_0 such that $U \cap Z = \{1\}$. Then U is isomorphic to $\pi(U)$, which is a subgroup of finite index in Δ_0 and hence in Δ.

Now take $\Delta_1 = \pi(U) \simeq U$. Observe that U is Zariski dense in the connected, simply connected, $\mathbb{Q}$-simple algebraic group $\widetilde{G}$. In addition, U is finitely generated and inside $\widetilde{G}_\mathbb{Q}$. It follows that there is a finite set S of rational primes such that $U \leq \widetilde{G}_{\mathbb{Z}_S}$. All the conditions of Theorem 5.1 are now satisfied with U and $\widetilde{G}$ in place of Δ and G. Hence we deduce that the congruence completion of U is an open subgroup of

$$G_S = \prod_{p \notin S} G_{\mathbb{Z}_p}.$$

This open subgroup projects onto all but finitely many of the factors in the product G_S. So by enlarging S to some finite set S_1 we may ensure that the congruence completion of U maps onto $\prod_{p \notin S_1} G_{\mathbb{Z}_p}$. Since U is isomorphic to Δ_1, Theorem 1.1 follows.

6 Applications of Lubotzky's alternative

As noted in the introduction, Theorem 1.1 puts a substantial restriction on the finite images of a linear group in characteristic 0. First we need to introduce the finite simple groups of Lie type.

6.1 The finite simple groups of Lie type

For a detailed account of the material of this section, we refer to Carter's book [4].

The *untwisted* simple groups of Lie type are the groups $L = G_{\mathbb{F}_q}/Z$ where G is a simply connected Chevalley group defined over $\mathbb{Z}$, and Z is the centre of the group of rational points $G_{\mathbb{F}_q}$ over the finite field $\mathbb{F}_q$. The *type* of L is just the Lie type $\mathcal{X}$ of G.

The *twisted* simple groups arise as the fixed points L^σ of a specific automorphism σ (of order 2 or 3) of some untwisted simple group L. Such twisted Lie type simple groups are for example $\mathrm{PSU}_n(q)$. The (untwisted) type of L^σ is just the Lie type of L. For example, the untwisted Lie type of $\mathrm{PSU}_n(q)$ is A_{n-1}.

A finite group L is quasisimple if $L = [L, L]$ and $L/Z(L)$ is simple. Similarly to the isogenies described in Theorem 2.18, the quasisimple finite groups break up into families with the same simple quotient. The members of each family have the same simple quotient, say S, and there is a largest member of the family L, called *the universal cover* of S. All the other members of the family are the quotients L/A where $A \leq Z(S)$. The type (twisted or not) of a quasisimple group is the same as that of its simple quotient.

6.2 Refinements

Let us return to Corollary 4.3. Recall that G was a simply connected, simple linear algebraic group defined over an algebraic number field k and $\Gamma = G_{\mathcal{O}_S}$ for a ring of algebraic S-integers $\mathcal{O}_S$ of k . The group Γ then maps onto $G_{\mathbb{F}_{q(\mathfrak{p})}}$ for almost all $\mathfrak{p} \notin S$.

Proposition 6.1. *Assume in the above situation that G is absolutely simple. Then for almost all prime ideals $\mathfrak{p}$ outside S the reduction $G_{\mathbb{F}_{q(\mathfrak{p})}}$ of G modulo $\mathfrak{p}$ is a quasisimple finite group.*

Now from the description of the k-forms of G it follows that G splits over $\mathbb{F}_{q(\mathfrak{p})}$ if and only if some specific polynomials in $k[x]$ (depending only on G) splits completely in linear factors in the finite field $\mathbb{F}_{q(\mathfrak{p})}$. The **Chebotarev density Theorem** now implies that $G_{\mathbb{F}_{q(\mathfrak{p})}}$ is an **untwisted** quasisimple group for a positive proportion of the primes $\mathfrak{p}$ of k.

Next consider the situation of Theorem 1.1. There we have a $\mathbb{Q}$-simple algebraic group G such that all congruence images of $G_{\mathbb{Z}_S}$ occur as quotients of Δ_1. Now G may not be absolutely simple but in any case there is a finite Galois extension k of $\mathbb{Q}$ and an absolutely simple k-defined group H such that

$G = \mathcal{R}_{k/\mathbb{Q}}H$. Proposition 4.4 gives that for almost all rational primes p outside S

$$G_{\mathbb{F}_p} = \prod_{\mathfrak{p}|p} H_{\mathbb{F}_{q(\mathfrak{p})}},$$

where as before the product on the right is over all prime ideals $\mathfrak{p}$ of k dividing the (unramified) prime p. Note that the degree of $\mathbb{F}_{q(\mathfrak{p})}$ over $\mathbb{F}_p$ is bounded by $(k : \mathbb{Q})$.

Therefore, Theorem 1.1 in combination with Corollary 4.3 gives:

Corollary 6.2. *Suppose that $\Gamma < \mathrm{GL}_n(K)$ is a finitely generated linear group in characteristic 0 which is not virtually soluble. Then there is:*

- *a positive integer d,*

- *a Lie type $\mathfrak{X}$,*

- *for every prime p a finite field $\mathbb{F}_{p^f}$ of degree $f \leq d$ over $\mathbb{F}_p$ and a finite simple group $L(p^f)$ of Lie type over $\mathbb{F}_{p^f}$ whose untwisted type is $\mathfrak{X}$ (e.g. if $\mathfrak{X} = A_{n-1}$, then $L(p^f)$ is either $\mathrm{PSL}_n(p^f)$ or $\mathrm{PSU}_n(p^f)$), and*

- *a subgroup of finite index Γ_0 in Γ,*

such that Γ_0 maps onto $L(p^f)$ for almost all primes p. Moreover, for a positive proportion of these primes one has $f = 1$ and the group $L(p)$ is untwisted.

One consequence of this is that Γ cannot have polynomial subgroup growth because the Cartesian product $\prod_{p\ \mathrm{prime}} L(p)$ does not have polynomial subgroup growth, see [12, Ch. 5.2] for details.

The untwisted type $\mathfrak{X}$ of the simple groups $L(p)$ is not completely arbitrary: Let G be the simple algebraic group of type $\mathfrak{X}$ as stated in Theorem 2.18. Then G is an image of the connected component of the Zariski closure of Γ in $G_n(K)$.

There is one particular case when the group G is explicitly determined: when Γ is a subgroup of $\mathrm{GL}_2(\mathbb{C})$. Then the dimension of G is at most 4. On the other hand, from the classification in Theorem 2.18 it follows that the only simple algebraic group of dimension less than 8 is SL_2. Therefore, we obtain the following:

Proposition 6.3. *A finitely generated subgroup Γ of $\mathrm{GL}_2(\mathbb{C})$ which is not virtually soluble has a subgroup of finite index Γ_0 which maps onto $\mathrm{PSL}_2(p)$ for infinitely many, in fact for a positive proportion of all primes p.*

This result is used in [13] where the authors prove that any lattice Λ in $\mathrm{PSL}_2(\mathbb{C})$ has a collection $\{N_i\}_i$ of subgroups of finite index such that $\bigcap_i N_i = \{1\}$ and Λ has property τ with respect to $\{N_i\}_i$. As a corollary, the authors obtain that any hyperbolic 3-manifold has a co-final sequence of finite covers with positive infimal Heegaard gradient.

6.3 Normal subgroups of linear groups

A normal subgroup N of a finitely generated group of course does not need to be finitely generated. So it comes as no surprise that when this happens in linear groups we can put further restriction on the finite images of N.

Proposition 6.4. *Let Γ be a finitely generated linear group with a finitely generated normal subgroup Δ. Assume that Δ is not virtually soluble. Then there exist a number $C > 0$ and a Lie type $\mathfrak{X}$ with the following property: for infinitely many primes $p \in \mathbb{N}$, the group Δ has a normal Γ-invariant subgroup N with Δ/N isomorphic to a direct product of at most k copies of the untwisted finite simple group $L(p)$ of type $\mathfrak{X}$ over $\mathbb{F}_p$.*

Sketch of proof. Using similar arguments to those in the proof of Theorem 1.1, we can reduce to the case when $\Gamma \leq \mathrm{GL}_d(\mathbb{Q})$ for some integer d, and Δ is Zariski dense in some absolutely simply connected, semisimple algebraic group $G \leq \mathrm{GL}_d$ defined over $\mathbb{Q}$ with isomorphic simple factors. Moreover, we have $\Gamma \leq \mathrm{GL}_d(\mathbb{Z}_S)$ for some finite set of rational primes S.

Let t be the number of simple factors of G.

As before for a rational prime $p \notin S$, let π_p be the homomorphism $\mathrm{GL}_d(\mathbb{Z}_S) \to \mathrm{GL}_d(\mathbb{F}_p)$ obtained by reducing $\mathbb{Z}_S$ mod p.

From Theorem 5.1 we deduce that for all but finitely many primes p outside S one has $\pi_p(\Delta) = G_{\mathbb{F}_p} = \pi_p(G_{\mathbb{Z}_S})$. Let $M_p = \ker \pi_p$ and $N_p = \Delta \cap M_p$. Then $\Delta/N_p \simeq G_{\mathbb{F}_p}$ is a central product of at most t quasisimple groups of the same Lie type as the factors of G. Also for infinitely many primes p, these factors are untwisted quasisimple groups.

Now the only thing remaining is to observe that M_p is normal in $\mathrm{GL}_d(\mathbb{Z}_S)$ and therefore $N_p = \Delta \cap M_p$ is invariant under Γ. Hence $G_{\mathbb{F}_p}/Z(G_{\mathbb{F}_p})$ is the required Γ-invariant quotient of Δ. $\square$

As suggested by Lubotzky, Proposition 6.4 may be relevant in the following open problem:

Conjecture 6.5 (Formanek–Zelmanov). *Let $n > 2$ and consider $\mathrm{Aut}(F_n)$, the automorphism of the free group on n free generators. If ρ is a complex linear representation of $\mathrm{Aut}(F_n)$, then $\rho(\mathrm{Inn}(F_n))$ is virtually soluble, where $\mathrm{Inn}(F_n)$ is the subgroup of inner automorphism of F_n.*

6.4 Representations, sieves and expanders

Knowledge of (sufficiently many of) the finite images of finitely generated linear group Γ has had many uses. Below we list in brief some of these applications without much detail. The interested reader is invited to look up the original articles.

- In [9] Theorem 1.1 is applied to obtain information about the collection of degrees of irreducible representations of a finitely generated linear group Γ.

- In [2] the authors use strong approximation to obtain sufficient conditions when the images of a Zariski dense finitely generated subgroup of $SL_n(\mathbb{Z})$ produce a family of expanding Cayley graphs for the groups $SL_n(\mathbb{F}_p)$. These conditions have been verified recently in [3] and [17].

- In [10] it is proved that if $\Gamma \leq SL_n(\mathbb{Z})$ is a finitely generated subgroup which maps onto $SL_n(\mathbb{F}_p)$ for some prime $p \neq 2, 3, n$, then in fact Γ maps onto $SL_n(\mathbb{F}_p)$ for almost all primes p. This is the content of Exercise 8.19.

- In [11] it is shown that the profinite completion of a finitely generated linear group over a field of characteristic different from 2 or 3 which is not virtually soluble cannot have finitely generated Sylow subgroups.

- In [18] the strong approximation theorem is used to prove that an arithmetic group Γ with GCSP which is a lattice in a higher rank semisimple Lie group must always contain a free group which is dense in the profinite topology of Γ.

7   The Nori–Weisfeiler theorem

Our sketch of the proof of Theorem 5.1 follows the argument in [12, Window 9].

Suppose that $\Gamma \leq G_{\mathbb{Z}_S}$ is Zariski dense in the simply connected, $\mathbb{Q}$-simple algebraic group G. Now G may not be absolutely simple, but in any case there is a number field k and an absolutely simple group H defined over k such that $G = \mathcal{R}_{k/\mathbb{Q}}(H)$. We have $G_{\mathbb{Q}} = H_k$, and for each prime p

$$G_{\mathbb{Z}_p} = \prod_j H_{\mathcal{O}_{\mathfrak{p}_j}},$$

where $p\mathcal{O} = \prod_j \mathfrak{p}_j^{e_j}$ is the factorisation of the principal ideal (p) in $\mathcal{O}$. This means that $k \otimes \mathbb{Q}_p = \prod_j k_{\mathfrak{p}_j}$.

From now on assume that the prime p is unramified in k, i.e. all $e_j = 1$. In addition, assume that G has good reduction mod p. This holds for all but finitely many rational primes p (see Proposition 4.1).

Since $L(G)$ is $\mathbb{Q}$-defined, we have that $L(G)_{\mathbb{Q}_p} = L(G) \otimes \mathbb{Q}_p$. Therefore, $L(G)_{\mathbb{Q}_p} = \prod_j L(H)_{k_{\mathfrak{p}_j}}$. Similarly, since p is unramified

$$L(G)_{\mathbb{F}_p} = \prod_j L(H)_{\mathcal{O}/\mathfrak{p}_j} \quad \text{and} \tag{7.1}$$

$$G_{\mathbb{F}_p} = \prod_j H_{\mathcal{O}/\mathfrak{p}_j}.$$

The group H is absolutely simple so for almost all primes p the Lie algebras $L(H)_{\mathcal{O}/\mathfrak{p}_j}$ are simple and the groups $H_{\mathcal{O}/\mathfrak{p}_j}$ are quasisimple.

Step 1: Let D_p be the closure of Δ in the p-adic analytic group $G_{\mathbb{Q}_p}$. Since Δ is Zariski-dense in G, then by Proposition 2.12 the Lie algebra of D is an ideal of the Lie algebra $L(G)_{\mathbb{Q}_p}$ of $G_{\mathbb{Q}_p}$. But $\Delta \leq G_{\mathbb{Q}}$, so the Lie algebra $L(D_p)$ is defined over $\mathbb{Q}$. Hence the projections of $L(D_p)$ in each of the factors $L(H)_{k_{\mathfrak{p}_j}}$ of $L(G)_{\mathbb{Q}_p}$ are isomorphic. So for almost all primes p we have $L(D_p) = L(G)_{\mathbb{Q}_p}$ which means that D_p is an open subgroup of $G_{\mathbb{Q}_p}$ for almost every p (see Proposition 2.13). In fact, since we are assuming $p \notin S$, we have $\Delta \subset G_{\mathbb{Z}_p}$, and so D_p is an open subgroup of the compact open subgroup $G_{\mathbb{Z}_p}$.

Next we want to prove that for almost all primes p, our group Δ is dense in $G_{\mathbb{Z}_p}$.

Step 2: For almost all primes, the Frattini subgroup of $G_{\mathbb{Z}_p}$ is contained in the kernel of $G_{\mathbb{Z}_p} \to G_{\mathbb{F}_p}$. It follows that a subgroup Δ is dense in $G_{\mathbb{Z}_p}$ if and only if Δ maps onto $G_{\mathbb{F}_p}$. This is proved in [12, Window 9, Proposition 7] using the structure of the finite images of the p-adic analytic group $G_{\mathbb{Z}_p}$.

Step 3: We shall prove that $D_p = G_{\mathbb{Z}_p}$ for almost all primes p. By Step 2 it is enough to show that Δ maps onto $G_{\mathbb{F}_p}$ for almost all primes p.

Let π_p be the projection of $G_{\mathbb{Z}_p}$ onto $G_{\mathbb{F}_p}$ and further let π_j and τ_j be the projections of $G_{\mathbb{Z}_p}$ and $L(G)_{\mathbb{Z}_p}$ onto their direct factors $H_{\mathbb{O}/\mathfrak{p}_j}$ and $L(H)_{\mathbb{O}/\mathfrak{p}_j}$ respectively.

At this stage we need the following:

Proposition 7.1. *Let Γ be a subgroup of $G_{\mathbb{F}_p}$ such that:*

(a) for all j, the image $\pi_j(\Gamma)$ of Γ in $H_{\mathbb{O}/\mathfrak{p}_j}$ has order divisible by p, and

(b) every subspace of $L(G)_{\mathbb{F}_p}$ invariant under Γ is an ideal.

Then provided p is sufficiently large compared to $\dim G$ we have $\Gamma = G_{\mathbb{F}_p}$.

Let us check that the conditions (a) and (b) above are satisfied for the group $\pi_p(\Delta) \leq G_{\mathbb{F}_p}$, for almost all primes p.

Suppose that (a) fails for a set A of infinitely many primes. Then there is $j = j_p$ such that $\pi_{j_p}(\Delta)$ has order coprime to p and so is a completely reducible subgroup of $\mathrm{GL}_n(\mathbb{F}_p)$, where n depends only on G and not on p. A variation of Jordan's theorem [8] then says that there is a number $f = f(n)$ such that $\pi_{j,p}(\Gamma)$ has an abelian subgroup of index at most f.

Since the set A of rational primes is infinite, we have

$$G_{\mathbb{Z}_S} \cap \bigcap_{p \in A} \ker \pi_{j_p} = \{1\}$$

This implies that Δ itself is virtually abelian (it is finitely generated so it has only finitely many subgroups of index at most $f(n)$). But Δ is Zariski-dense in the $\mathbb{Q}$-simple algebraic group G: contradiction.

So condition (a) of Proposition 7.1 holds for almost all primes.

Condition (b) is immediate: H is absolutely simple and so for almost all primes, each of the $L(H)_{\mathbb{O}/\mathfrak{p}_j}$ is a simple module for $H_{\mathbb{O}/\mathfrak{p}_j}$. Since Δ is Zariski-dense in

H_k, the group $\mathrm{Ad}(\Delta)$ spans $\mathrm{End}_k L(H)_k$ and so for almost all primes, $\mathrm{Ad}(\pi_j(\Delta))$ spans $\mathrm{End}_{\mathcal{O}/\mathfrak{p}_j} L(H)_{\mathcal{O}/\mathfrak{p}_j}$. This means that each summand $L(H)_{\mathcal{O}/\mathfrak{p}_j}$ of $L(G)_{\mathbb{F}_p}$ is a simple module for $\pi_p(\Delta)$. So the decomposition of $L(G)_{\mathbb{F}_p}$ into minimal Lie ideals is also a decomposition into irreducible $\mathbb{F}_p\pi_p(\Delta)$-modules. So every irreducible module for $\pi_p(\Delta)$ in $L(G)_{\mathbb{F}_p}$ is an ideal, proving that (b) holds.

Step 4 We now know that the closure $\overline{\Delta}$ of Δ in $G_{\widehat{\mathbb{Z}}_S} = \prod_{p \notin S} G_{\mathbb{Z}_p}$ projects onto all but finitely many of the factors $G_{\mathbb{Z}_p}$. Now it is easy to show (see Exercise 8.16) that in this case $\overline{\Delta}$ contains their Cartesian product. Combined with Step 1 (which says that $\overline{\Delta}$ projects onto an open subgroup in each of the remaining factors) we easily see that $\overline{\Delta}$ is open in $G_{\widehat{\mathbb{Z}}_S}$.

7.1   Unipotently generated subgroups of algebraic groups over finite fields

There are now at least three different proofs of Proposition 7.1. One is by Matthews, Vaserstein and Weisfeiler [14]; it uses the classification of the finite simple groups to deduce properties of a proper subgroup of $G_{\mathbb{F}_p} \le \mathrm{GL}_n(\mathbb{F}_p)$ which are incompatible with (a) and (b).

There is also a proof using logic by Hrushovski and Pillay [6].

We shall focus on the original proof by Nori [15]. It studies unipotently generated algebraic groups and their Lie algebras in large finite characteristic p. This is motivated by the construction of the Chevalley groups described in Section 2.5. Recall that the adjoint Chevalley group $\overline{G}$ is generated by certain automorphisms $\exp(\mathrm{ad}(x))$ for certain ad-nilpotent elements x of the Lie algebra of G. If we fix such an element x, then the set

$$\{\exp(\mathrm{ad}(tx)) \mid t \in K\}$$

is a unipotent subgroup of $\overline{G}$ and is isomorphic to $\mathbb{G}_+$.

Nori generalises this situation in two directions: he proves an analogue of this not just for algebric groups but for Zariski-dense subgroups of $\mathrm{GL}_n(\overline{\mathbb{F}_p})$ and, secondly, he does this not just in the algebraic closure $\overline{\mathbb{F}_p}$ of $\mathbb{F}_p$ but in the finite field $\mathbb{F}_p$ (provided p is large enough compared to n).

The details are as follows:

For a group $\Gamma \le \mathrm{GL}_n(\mathbb{F}_p)$, let Γ^+ be the subgroup generated by its unipotent elements. When $p \ge n$ these are just the elements of order p in Γ. Similarly, for an algebraic group $G \le \mathrm{GL}_n(K)$, let G^+ be the subgroup generated by its unipotent elements.

Now for an element $g \in \mathrm{GL}_n(\mathbb{F}_p)$ of order p, let X_g be the unipotent 1-dimensional algebraic group over $\overline{\mathbb{F}_p}$ generated by g. In other words, define

$$X_g = \left\{ g^t := \sum_{i=0}^{p} \binom{t}{i}(g-1)^i \;\middle|\; t \in \overline{\mathbb{F}_p} \right\},$$

where $\overline{\mathbb{F}}_p$ is the algebraic closure of $\mathbb{F}_p$. Note that X_g is defined over $\mathbb{F}_p$ and is isomorphic to the additive group of the field $\overline{\mathbb{F}}_p$.

Now, given $\Gamma \leq \mathrm{GL}_n(\mathbb{F}_p)$, define the algebraic group $T = T(\Gamma)$ as

$$T = \langle X_g \mid \forall g \in \Gamma, g^p = 1 \rangle \leq \mathrm{GL}_n(\overline{\mathbb{F}}_p).$$

Recall that the subgroup generated by a collection of closed connected subgroups is closed and connected, so T is indeed a connected algebraic group. Observe that since X_g is the smallest connected algebraic group containing g and $g \in G_{\mathbb{F}_p}$, it follows that $X_g \leq G$ and hence $T \leq G$.

Nori's main result is that in the above setting we have

$$\Gamma^+ = (T_{\mathbb{F}_p})^+,$$

provided p is large enough compared to n.

Now, it is known that for large primes p one has

$$(T_{\mathbb{F}_p})^+ = T_{\mathbb{F}_p}.$$

So Γ^+ is the group of $\mathbb{F}_p$-rational points of the connected algebraic group T.

Now, suppose that condition (b) of Proposition 7.1 holds. Clearly, Γ normalises the algebraic group $T \leq G$ since $(X_g)^\gamma = X_{g^\gamma}$ for any $\gamma, g \in \Gamma$ with $g^p = 1$. Therefore, the Lie algebra $L(T) \leq L(G)$ of T is normalised by Γ.

It follows that the subspace $L(T)_{\mathbb{F}_p}$ of $L(G)_{\mathbb{F}_p}$ is invariant under Γ and so by assumption (b) of Proposition 7.1 it is an ideal of $L(G)_{\mathbb{F}_p}$. Not only that, $L(T)$ is defined over $\mathbb{F}_p$ and so its projections on the direct factors of $L(G)$ are isomorphic. In the same way as in Step 1 above we deduce that $L(T) = L(G)$ and since both G and $T \leq G$ are connected we have $T = G$. So

$$\Gamma \geq \Gamma^+ = T_{\mathbb{F}_p} = G_{\mathbb{F}_p} \geq \Gamma$$

giving that $\Gamma = G_{\mathbb{F}_p}$ as required.

8 Exercises

Exercise 8.1. Show that every open set in K^n can be regarded as closed affine set in some K^m, $m \geq n$.

Exercise 8.2. Prove that $\dim V$ for an irreducible affine variety V is the largest d such that we can find a chain $\emptyset \neq V_1 \subset V_2 \subset \cdots V_d \subset V$ of distinct irreducible closed subvarieties V_i in V. You may use any of the equivalent definitions of $\dim V$ in Section 2.1.

Exercise 8.3 (This is Proposition 2.2). Show that each affine variety is a compact topological space and that in fact it satisfies the descending chain condition on closed subsets.

A subset $X \subset V$ of an affine variety V is *constructible* if it can be obtained from the open or closed subsets of V by a finite process of forming unions and intersections. A theorem of Chevalley says that an image of a constructible set under a morphism of varieties is constructible.

Exercise 8.4 (see [20, Lemma 14.10]). Prove that a constructible (abstract) subgroup H of a linear algebraic group G is in fact closed, and so is algebraic. Deduce with Chevalley's theorem that an image of an algebraic group under a homomorphism is an algebraic group.

Exercise 8.5 (see [20, Lemma 14.14]). Let G be a linear algebraic group and $(X_i)_{i \in I}$ be a family of constructible irreducible subsets of G each containing the identity. Show that X_i together generate a closed irreducible subgroup of G. Hence deduce that if G is connected, then the derived subgroup $G' = \langle [x, y] \mid x, y \in G \rangle$ is both closed and connected.

Exercise 8.6. Suppose that k/k_0 is a finite extension of fields and $H = \mathcal{R}_{k/k_0}(G)$. Show that H is K-isomorphic to

$$G^{\sigma_1} \times G^{\sigma_2} \times \cdots \times G^{\sigma_d},$$

where σ_i are all the embeddings of k in K which fix the elements of k_0, and G^{σ_i} is the algebraic group defined by the ideal I^{σ_i} where the ideal I defines $G = V(I)$ as a variety in $M_n(K)$. *Hint*: use the map λ on page 70 and the isomorphism (2.1).

Exercise 8.7. Let G be the multiplicative group of norm one quaternions defined over $\mathbb{Q}$. For example, we can take G in its left regular representation

$$G = \left\{ \begin{pmatrix} a & -b & -c & -d \\ b & a & d & -c \\ c & -d & a & b \\ d & c & -b & a \end{pmatrix} \mid a^2 + b^2 + c^2 + d^2 = 1 \right\}$$

Show that G is $\mathbb{Q}(i)$-isomorphic to SL_2 but it is not $\mathbb{Q}$-isomorphic to it. *Hint*: send the 4×4 matrix with first column a, b, c, d as above to

$$\begin{pmatrix} a + ib & -c + id \\ c + id & a - ib \end{pmatrix}.$$

Exercise 8.8. Show that if $G = \mathrm{SL}_n(K)$, then $L(G) = \mathrm{sl}_n(K)$, the Lie algebra of matrices of trace 0 in $M_n(K)$.

Exercise 8.9. Show that $\Gamma = \mathrm{SL}_2(\mathbb{Z})$ does not have the generalised congruence subgroup property. You may use that Γ has a non-abelian free subgroup of finite index.

Exercise 8.10. Show that $\mathrm{SL}_n(\mathbb{Z})$ has the strong approximation property. (*Hint*: use the fact that for a finite ring R the group $\mathrm{SL}_n(R)$ is generated by elementary matrices.)

Exercise 8.11. Show that $\mathrm{PGL}_2(\mathbb{Z})$ fails to have the strong approximation property (as an arithmetic subgroup of $G = \mathrm{PGL}_2$).

Exercise 8.12. Show that if a connected linear algebraic group G is not soluble, then it maps onto a simple algebraic group. (*Hint*: let $M = \mathrm{Rad}\, G$ be the soluble radical of G. Then G/M is semisimple.)

Exercise 8.13. Suppose that Γ is a Zariski-dense subgroup of a connected algebraic group G and that Δ is a subgroup of finite index in Γ. Show that Δ is also Zariski-dense in G.

Exercise 8.14. Suppose that $G \le \mathrm{GL}_n(K)$ is a connected algebraic group with a normal subgroup N, which preserves a 1-dimensional subspace $\langle v \rangle$. Show that either N acts as scalars or else G stabilises a non-trivial subspace of K^n.

Exercise 8.15. Show that a connected soluble algebraic group $G \le \mathrm{GL}_n(K)$ has a common eigenvector. Deduce that G is triangularisable and hence prove Theorem 1.2. (*Hint*: use Exercise 8.14 with G' in place of N.)

Exercise 8.16. Suppose that L is a closed subgroup of $K = \prod_{p \in A} G_{\mathbb{Z}_p}$ for some set A of primes, where G is a connected, simply connected, $\mathbb{Q}$-simple algebraic group.

(a) Show that if p is sufficiently large, then if L maps onto the direct factor $G_{\mathbb{Z}_p}$ of K it in fact contains it.

(b) On the other hand, if A is finite set of primes and L maps onto an open subgroup of each factor $G_{\mathbb{Z}_p}$ of K, show that then L is an open subgroup of K.

Exercise 8.17. Show that for any algebraic group G in characteristic 0 the group G^+ generated by its unipotent elements is connected. (*Hint*: use Exercise 8.5)

Exercise 8.18. Using Theorems 2.25 and 2.18 show that if a connected algebraic group consists of semisimple elements, then it is a torus. (*Hint*: a non-trivial semisimple group contains a copy of SL_2 or PSL_2.)

Exercise 8.19 (see [10]). Let $n > 1$ be an integer. Show using the strong approximation theorem that there is a finite set A of rational primes with the following property: if $S \subseteq \mathrm{SL}_n(\mathbb{Z})$ is a subset whose image generates $\mathrm{SL}_n(\mathbb{F}_p)$ for some prime $p \notin A$, then for almost all primes q the image of S in $\mathrm{SL}_n(\mathbb{F}_q)$ generates $\mathrm{SL}_n(\mathbb{F}_q)$. Generalise this to any connected, simply connected, absolutely simple group G defined over $\mathbb{Z}$.

References for Chapter II

[1] M. F. Atiyah and I. G. MacDonald. *Introduction to commutative algebra*, Addison-Wesley series in Mathematics, 1969.

[2] J. Bourgain, A. Gamburd and P. Sarnak. Affine linear sieve, expanders, and sum-product. *Inventiones Math.*, 179(3): 559–644, 2010.

[3] E. Breuillard, B. Green and T. Tao. Linear approximate subgroups, preprint http://arxiv.org/abs/1001.4570.

[4] R. W. Carter. *Simple groups of Lie type*. London and New York, Wiley, 1972.

[5] R. W. Carter, I. G. Macdonald and G. B. Segal. *Lectures on Lie groups and Lie algebras*, LMS Student Texts 32, Cambridge University Press, Cambridge, 1995.

[6] E. Hrushovski and E. Pillay. Definable subgroups of algebraic groups over finite fields. *J. reine angew. Math.*, 462: 69–91, 1995.

[7] J. Humphreys. *Linear algebraic groups, Graduate Texts in Mathematics* No. 21, Springer-Verlag, 1975.

[8] C. Jordan. Mémoire sur les equations différentielles linéaires à intégrale algébrique, *J. reine angew. Math.*, 84: 89–215, 1878.

[9] M. W. Liebeck, D. Segal and A. Shalev. The density of representation degrees. Forthcoming.

[10] A. Lubotzky. One for almost all: generation of SL(n, p) by subsets of SL(n, Z). In T. Y. Lam and A. R. Magid (eds.), *Algebra, K-. Theory, Groups and Education, Contemp. Math.*, 243, Amer. Math. Soc., Providence, RI, 1999, pp. 125–128.

[11] A. Lubotzky. On finite index subgroups of linear groups. *Bull. London Math. Soc.*, 19: 325–328, 1987.

[12] A. Lubotzky and D. Segal. *Subgroup growth*, Birkhäuser, Basel, 2003.

[13] D. D. Long, A. Lubotzky and A. W. Reid. Heegaard genus and Property τ for hyperbolic 3-manifolds. *J. Topol.*, 1(1): 152–158, 2008.

[14] C. R. Matthews, L. N. Vaserstein and B. Weisfeiler. Congruence properties of Zariski-dense subgroups. *Proc. London Math. Soc.*, 48: 514–532, 1984.

[15] M. Nori. On subgroups of $GL_n(\mathbb{F}_p)$. *Invent. Math.*, 88: 257–275, 1987.

[16] V. Platonov and A. Rapinchuk. *Algebraic groups and number theory.* Academic Press, 1994.

[17] L. Pyber and E. Szabo. Growth in finite simple groups of Lie type, preprint arXiv:1001.4556.

[18] G. A. Soifer and T. N. Venkataramana. Finitely generated profinitely dense free groups in higher rank semi-simple groups. *Transform. Groups*, 5(1): 93–100, 2000.

[19] J. Tits. Classification of algebraic semisimple groups, in A. Borel and G. Mostow (eds.), *Algebraic groups and discontinuous subgroups. Proc. Symposia Pure Math.*, 9, Amer. Math. Soc., 1966.

[20] B. Wehrfritz. *Infinite linear groups.* Springer-Verlag, 1973.

[21] B. Weisfeiler. Strong approximation for Zariski-dense subgroups of semisimple algebraic groups. *Annals of Math.*, 120: 271–315, 1984.

Chapter III

A newcomer's guide to zeta functions of groups and rings

by Christopher Voll

1 Introduction

1.1 Zeta functions of groups

A finitely generated group G has only finitely many subgroups of each finite index. Denoting, for $m \in \mathbb{N}$, by $a_m = a_m(G)$ the number of subgroups of index m in G, we define the *subgroup zeta function* of G as the formal Dirichlet series encoding these numbers

$$\zeta_G(s) := \sum_{m=1}^{\infty} a_m m^{-s} = \sum_{H \leq_f G} |G : H|^{-s}. \tag{1.1}$$

Here s is a complex variable.

Over the last few decades, zeta functions have been developed into a major tool in the study of groups with 'slow' subgroup growth. Zeta functions of groups share a number of key properties with classical zeta functions in number theory, algebraic geometry and combinatorics. These include, for instance, the Dedekind zeta functions of number fields, the Hasse–Weil zeta functions associated to algebraic varieties over finite fields and generating functions enumerating integral points of rational polyhedral cones. An important feature in the theory of zeta functions of groups is the fact that much of it may be 'linearised' by passing to suitable Lie rings. This is one motivation for studying zeta functions of rings (of finite additive rank) counting subrings of finite index. The idea to study groups via associated Lie rings or algebras is, of course, a very general one. It also lies at the heart of the subject matter in Chapter I.

In the current chapter, we present a number of key features and results in the theory of zeta functions of groups and rings, including Euler products, analytic properties, rationality results and local functional equations. We aim to introduce the reader to some of the key techniques used in the area, including tools from p-adic integration, algebraic geometry and combinatorics, exploring some of the connections with classical zeta functions mentioned above.

We say that a finitely generated group G has *polynomial subgroup growth* (PSG) if the sequence

$$s_m := s_m(G) := \sum_{i \leq m} a_i$$

is bounded by a polynomial in m. It is a fundamental fact from the theory of Dirichlet series that the formal series (1.1) converges on some right-half plane of complex numbers if and only if G has PSG. For such a group, the *abscissa of convergence* $\alpha(G)$ of (1.1), i.e. the infimum of all $\alpha \in \mathbb{R}$ such that $\zeta_G(s)$ converges on $\{s \in \mathbb{C} \mid \mathrm{Re}(s) > \alpha\}$, determines the exact degree of polynomial subgroup growth of G: we have $s_m(G) = O(1 + m^{\alpha(G)+\varepsilon})$ for every $\varepsilon \in \mathbb{R}_{>0}$. A classical reference on Dirichlet series is [23]. See [1, Chapter 11] for a more modern treatment.

The degree of subgroup growth of any PSG group G is the same as the degree of subgroup growth of $G/R(G)$, where $R(G) := \bigcap_{N \triangleleft_f G} N$, the finite residual of G, is the intersection of the group's normal subgroups of finite index. In studying subgroup growth, we may thus assume without loss of generality that the group G is residually finite, i.e. that its finite residual is trivial. Finitely generated, residually finite groups of polynomial subgroup growth have been characterised as the virtually soluble groups of finite rank; cf. [37]. The importance of the arithmetic function $m \mapsto a_m(G)$ in this context is already emphasised in [43].

The class of finitely generated, residually finite PSG groups includes the class of finitely generated, torsion-free nilpotent (or $\mathcal{T}$-)groups. It was this class of groups for which zeta functions were first introduced as a means to study asymptotic and arithmetic aspects of the subgroup growth function; cf. [22].

Let G be a $\mathcal{T}$-group. It is not difficult to see that, owing to the nilpotency of G, the zeta function $\zeta_G(s)$ has an Euler factorisation

$$\zeta_G(s) = \prod_{p \text{ prime}} \zeta_{G,p}(s) \tag{1.2}$$

into local (or Euler) factors $\zeta_{G,p}(s) := \sum_{i=0}^{\infty} a_{p^i} p^{-is}$, indexed by the primes p, enumerating subgroups of p-power index. These may also be interpreted as zeta functions of the pro-p completions $\hat{G}_p$. Equation (1.2) generalises the familiar Euler product for the Riemann zeta function

$$\zeta(s) := \sum_{m=1}^{\infty} m^{-s} = \prod_{p \text{ prime}} \zeta_p(s), \tag{1.3}$$

where $\zeta_p(s) := 1/(1 - p^{-s})$. While (1.3) reflects the Fundamental Theorem of Arithmetic that every positive integer can be written as the product of prime powers in an essentially unique way, the identity (1.2) reflects the fact that every *finite* nilpotent group is the direct product of its Sylow p-subgroups. In fact, $\zeta(s)$ is the zeta function of the infinite cyclic group, making (1.3) a special case of (1.2). Indeed, it is well known that, for all $n \in \mathbb{N}$, there is a unique subgroup of index n in $\mathbb{Z}$, namely $n\mathbb{Z}$. It is instructive to see how this generalises to abelian groups of higher rank.

Example 1.1. For $n \in \mathbb{N}$, let $\mathbb{Z}^n$ be the free abelian group of rank n. Then

$$\zeta_{\mathbb{Z}^n}(s) = \zeta(s)\zeta(s-1)\cdots\zeta(s-(n-1)). \tag{1.4}$$

The monograph [39] contains no fewer than five proofs of this beautiful formula. We will add another, new one, in Section 2.5. We observe that (1.4) allows us to give precise asymptotic information about the numbers $s_m(\mathbb{Z}^n)$ of subgroups of index at most m in $\mathbb{Z}^n$. Indeed, one can deduce from (1.4) that

$$s_m(\mathbb{Z}^n) \sim n^{-1}\zeta(n)\zeta(n-1)\ldots\zeta(2)m^n \quad \text{as } m \to \infty.$$

Here '$f(m) \sim g(m)$ as $m \to \infty$' means that $\lim_{m\to\infty} f(m)/g(m) = 1$. For example, using the identity $\zeta(2) = \pi^2/6$, we see that

$$s_m(\mathbb{Z}^2) \sim (\pi^2/12)\, m^2 \quad \text{as } m \to \infty.$$

1.2 Zeta functions of rings

By a ring we mean an abelian group of finite rank, written additively, carrying a bi-additive product, not necessarily commutative or associative. We do not assume the existence of a multiplicative identity. Given a ring L, its *subring zeta function* is defined as the Dirichlet generating series

$$\zeta_L(s) = \sum_{m=1}^{\infty} b_m m^{-s} = \sum_{H \leq_f L} |L : H|^{-s},$$

where, for $m \in \mathbb{N}$, we denote by $b_m = b_m(L)$ the number of subrings of index m in L and s is a complex variable. (Note that, even if L is a ring with unity, we do not postulate that a subring should contain the unity element.) By properties of the underlying additive group of L – essentially the Chinese remainder theorem – this zeta function also satisfies an Euler product decomposition

$$\zeta_L(s) = \prod_{p \text{ prime}} \zeta_{L,p}(s) \tag{1.5}$$

into Euler factors $\zeta_{L,p}(s) := \sum_{i=0}^{\infty} b_{p^i} p^{-is}$, enumerating subrings of finite p-power index. It is worth pointing out that, for each prime p, the Euler factor $\zeta_{L,p}(s)$ is the zeta function of the $\mathbb{Z}_p$-algebra $L_p := L \otimes_{\mathbb{Z}} \mathbb{Z}_p$, enumerating subalgebras of finite index. Here $\mathbb{Z}_p$ is the ring of p-adic integers. We remark that the subring zeta function of a ring of additive torsion-free rank n always converges on $\{s \in \mathbb{C} \mid \mathrm{Re}(s) > n\}$, as the subring growth of L is bounded by the subgroup growth of $\mathbb{Z}^n$. The degree of the latter is n; cf. Example 1.1.

We will see in Section 1.3 that the study of subgroup zeta functions of nilpotent groups may – at least to a certain extent – be reduced to the study of subring zeta functions of (nilpotent Lie) rings. A *Lie ring* is a finitely generated abelian group with a bi-additive product $[,]$ (called 'Lie bracket') satisfying the Jacobi identity

$$\forall x, y, z \in L : \quad [x, [y, z]] + [y, [z, x]] + [z, [x, y]] = 0$$

and, for all $x \in L$, $[x, x] = 0$.

Example 1.2. Let $\mathfrak{sl}_2(\mathbb{Z})$ be the Lie ring of traceless integral 2×2-matrices with Lie bracket $[x, y] := xy - yx$. It has a $\mathbb{Z}$-basis consisting of the matrices

$$e := \begin{pmatrix} 0 & 1 \\ 0 & 0 \end{pmatrix}, \quad f := \begin{pmatrix} 0 & 0 \\ 1 & 0 \end{pmatrix}, \quad h := \begin{pmatrix} 1 & 0 \\ 0 & -1 \end{pmatrix}.$$

They satisfy the relations $[h, e] = 2e$, $[h, f] = -2f$, $[e, f] = h$. A non-trivial computation shows that, for every odd prime p, we have

$$\zeta_{\mathfrak{sl}_2(\mathbb{Z}),p}(s) = \zeta_{\mathfrak{sl}_2(\mathbb{Z}_p)}(s) = \zeta_p(s)\zeta_p(s-1)\zeta_p(2s-1)\zeta_p(2s-2)\zeta_p(3s-1)^{-1},$$

whereas, for $p = 2$, we have

$$\zeta_{\mathfrak{sl}_2(\mathbb{Z}),2}(s) = \zeta_{\mathfrak{sl}_2(\mathbb{Z}_2)}(s) = \zeta_2(s)\zeta_2(s-1)\zeta_2(2s-1)\zeta_2(2s-2)(1+6 \cdot 2^{-2s} - 8 \cdot 2^{-3s}).$$

This was proved in [20]. We sketch an alternative proof in Section 2.6.

The lower central series of a Lie ring L is defined inductively via $\gamma_1(L) := L$, $\gamma_i(L) := [\gamma_{i-1}(L), L]$ for $i \geq 2$. We say that a Lie ring L is nilpotent of class c if $\gamma_{c+1}(L) = \{0\}$ but $\gamma_c(L) \neq \{0\}$. For example, a Lie ring is nilpotent of class 1 if and only if it is abelian, and nilpotent of class at most 2 if and only if the derived ring is central, i.e. if $L' := \gamma_2(L) \leq Z(L)$.

Example 1.3. The Heisenberg Lie ring

$$L = \langle x, y, z \mid [x, y] = z, [x, z] = [y, z] = 0 \rangle$$

has $\mathbb{Z}$-rank 3 and nilpotency class 2. It can be shown that

$$\zeta_L(s) = \zeta(s)\zeta(s-1)\zeta(2s-2)\zeta(2s-3)\zeta(3s-3)^{-1}. \tag{1.6}$$

This was first proved in [22]. We establish (1.6) in Proposition 2.12 and sketch another proof in Section 2.6.

1.3 Linearisation

Whilst it is possible to analyse the Euler factors of subgroup zeta functions of nilpotent groups directly (cf. [22, Section 2]), it is often useful to exploit the fact that the study of subgroup growth of nilpotent groups can be linearised, i.e. reduced to the study of subring growth of suitable (nilpotent Lie) rings associated to these groups. Let G be a $\mathcal{T}$-group. The Mal'cev correspondence assigns to G a $\mathbb{Q}$-Lie algebra $\mathcal{L} = \mathcal{L}(G)$ which contains a Lie subring $L = L(G)$ such that $\mathcal{L} = L \otimes_{\mathbb{Z}} \mathbb{Q}$. The dimension of $\mathcal{L}$ as a $\mathbb{Q}$-vector space, and thus the torsion-free rank of L as a $\mathbb{Z}$-module, coincides with the Hirsch length $h(G)$ of G, the number of infinite cyclic factors in a polycyclic series for G. It can also be shown that L is nilpotent of class c, where c is the nilpotency class of G. The ring has the property that, for almost all (i.e. all but finitely many) primes p, one has

$$\zeta_{G,p}(s) = \zeta_{L,p}(s). \tag{1.7}$$

See [22, Section 4] for details. The exclusion of a finite number of primes is a recurrent phenomenon in the theory of global zeta functions associated to groups and rings.

If G is nilpotent of class 1, i.e. abelian, there is of course nothing to do: we choose $(L,+) = (G,\cdot)$, with trivial ring structure. If G is nilpotent of class 2, i.e. if $G' \le Z(G)$, we may choose, as an alternative to the Lie ring provided by the Mal'cev correspondence, the Lie ring

$$L := Z(G) \oplus G/Z(G), \tag{1.8}$$

with Lie bracket induced from taking commutators in G. It satisfies the identities (1.7) for all primes, and hence $\zeta_G(s) = \zeta_L(s)$ in this case.

We illustrate the passage from nilpotent groups to nilpotent Lie rings with an important and prototypical example and some of its generalisations.

Example 1.4. The group

$$G := \begin{pmatrix} 1 & \mathbb{Z} & \mathbb{Z} \\ 0 & 1 & \mathbb{Z} \\ 0 & 0 & 1 \end{pmatrix}$$

is called the discrete Heisenberg group of 3×3-upper-unitriangular matrices. It can easily be seen to be of Hirsch length 3 and nilpotent of class 2. In fact, its centre $Z(G)$ coincides with the derived group G', which is the infinite cyclic subgroup generated by the matrix

$$\begin{pmatrix} 1 & 0 & 1 \\ 0 & 1 & 0 \\ 0 & 0 & 1 \end{pmatrix}.$$

It is not hard to see that the Lie ring L constructed in (1.8) is isomorphic to the Heisenberg Lie ring from Example 1.3.

Example 1.5. The Heisenberg group has many aspects that may be generalised. For instance, it is the free nilpotent group of nilpotency class 2 on two generators. In general, given integers $c, d \geq 2$, the free nilpotent group $F_{c,d}$ on d generators and nilpotency class c may be defined as the quotient

$$F_{c,d} := F_d / \gamma_{c+1}(F_d)$$

of the free group F_d on d letters by the $(c+1)$th term of its lower central series. For $d \geq 2$, the group $F_{2,d}$, for example, has a presentation

$$F_{2,d} = \langle x_1, \ldots, x_d, y_{12}, y_{13}, \ldots, y_{d-1\,d} \mid$$
$$\forall i < j : [x_i, x_j] = y_{ij}, \forall k \, \forall i < j : [x_k, y_{ij}] = 1 \rangle.$$

The associated Lie rings $L_{2,d}$ have analogous presentations.

Computing explicit formulae for subgroup zeta functions of groups is in general very difficult, even if the groups have quite a transparent structure. For the zeta functions $\zeta_{F_{c,d}}(s)$, explicit formulae are only known for the cases $(1, d)$, $d \in \mathbb{N}$ (cf. Example 1.1), and $(c, d) \in \{(2, 2), (2, 3), (3, 2)\}$. For example

$$\zeta_{F_{2,3}}(s) = \zeta_{\mathbb{Z}^3}(s)\zeta(2s - 4)\zeta(2s - 5)\zeta(2s - 6)\zeta(3s - 6) \cdot$$
$$\zeta(3s - 7)\zeta(3s - 8)\zeta(4s - 8)^{-1} \prod_{p \text{ prime}} W_{2,3}(p, p^{-s}),$$

where

$$W_{2,3}(X, Y) = 1 + X^3 Y^2 + X^4 Y^2 + X^5 Y^2 - X^4 Y^3 - X^5 Y^3$$
$$- X^6 Y^3 - X^7 Y^4 - X^9 Y^4 - X^{10} Y^5 - X^{11} Y^5$$
$$- X^{12} Y^5 + X^{11} Y^6 + X^{12} Y^6 + X^{13} Y^6 + X^{16} Y^8;$$

see [21, 2.7.1].

We have seen that the theory of subgroup zeta functions of nilpotent groups can, to a great extent, be reduced to the study of the subring zeta functions of nilpotent Lie rings. It is worth recalling, however, that the theory of zeta functions of rings we are about to present applies to much more general rings.

1.4   Organisation of the chapter

In these notes we concentrate mainly on zeta functions associated with rings and nilpotent groups. In Section 2 we review some of the methods available to study local and global aspects of these functions, focussing on zeta functions of rings. By the linearisation results outlined in Section 1.3 these methods yield also theorems about zeta functions of $\mathcal{T}$-groups. We put particular emphasis on connections with the theory of linear homogeneous diophantine equations and on local functional equations.

Some of the manifold generalisations and variations of the concept of the zeta function of a nilpotent group or a ring are reviewed in Section 3. In Section 3.1

we briefly discuss ideal zeta functions of rings and normal zeta functions of nilpotent groups. These may be closest to the classical number-theoretic analogues of zeta functions of groups and rings. Section 3.2 gives an exposition of some recent developments in the theory of representation zeta functions of groups, focussing on finitely generated nilpotent, semisimple arithmetic and compact p-adic analytic groups. In Section 3.3 we point to a few variations of the counting problems discussed in these notes. They include subgroup zeta functions of compact p-adic analytic groups and an application of the theory of normal zeta functions of nilpotent groups to the enumeration of finite p-groups.

In Section 4 we present a collection of some major open questions and conjectures in the area.

There are a number of introductory and survey texts on zeta functions of groups. The texts [39, Chapters 15 and 16], [15] and [19, Chapter 9] provide further and complimentary reading.

We use the following notation:

$\mathbb{N}$	the set $\{1, 2, \dots\}$ of natural numbers
$I = \{i_1, \dots, i_l\}_<$	the finite set of natural numbers $i_1 < \cdots < i_l$
I_0	the set $I \cup \{0\}$ for $I \subseteq \mathbb{N}$
S_n	the symmetric group on n letters
M^t	the transpose of a matrix M
$\mathbb{Z}_p$	the ring of p-adic integers
$\mathbb{Q}_p$	the field of p-adic numbers
v_p	the p-adic valuation
$\lvert \ \rvert_p$	the p-adic absolute value,
	defined by $\lvert x \rvert_p := p^{-v_p(x)}$ for $x \in \mathbb{Q}_p$
$[\Lambda]$	the homothety class $\mathbb{Q}_p^* \Lambda$ of a (full) lattice Λ in $\mathbb{Q}_p^n$

Given a ring R, $r \in \mathbb{N}$, a set $\mathbf{f} \subseteq R[x_1, \dots, x_r] =: R[\mathbf{x}]$ of polynomials and a polynomial $g \in R[\mathbf{x}]$, we write $g\mathbf{f}$ for $\{gf \mid f \in \mathbf{f}\} \subseteq R[\mathbf{x}]$, and $(\mathbf{f}) \trianglelefteq R[\mathbf{x}]$ for the polynomial ideal generated by $\mathbf{f}$.

Acknowledgements. I am indebted to Mark Berman, Benjamin Klopsch and Alexander Stasinski, whose careful comments greatly improved these notes.

2 Local and global zeta functions of groups and rings

Let L be a ring of finite additive rank. By convention we will, in this section, use the term 'zeta function of L' to refer to the subring zeta function of L. Analogously, we refer to the subgroup zeta function of a group as its 'zeta function'.

By virtue of the Euler product (1.5), the problem of studying the zeta function $\zeta_L(s)$ is reduced to the problem of understanding the Euler factors $\zeta_{L,p}(s)$, p prime, and the analytic properties of their product. The following are natural questions. What do the local factors $\zeta_{L,p}(s)$ have in common? What is their structure? How do the local factors vary with the prime p? In the current section we explore some of the existing methods to analyse local and global zeta functions of rings and to address these questions.

2.1 Rationality and variation with the prime

In all the examples we have seen in Section 1 the local factors shared a number of features. In particular, they were all rational functions in the parameter p^{-s}. This is no coincidence:

Theorem 2.1 ([22, Theorem 3.5]). *Let p be a prime. The local zeta function $\zeta_{L,p}(s)$ is a rational function in p^{-s}, i.e. there is a rational function $W_p(Y) = P_p(Y)/Q_p(Y) \in \mathbb{Q}(Y)$ such that*

$$\zeta_{L,p}(s) = W_p(p^{-s}).$$

The proof of this theorem uses deep results from the theory of p-adic integration which we survey to some degree below.

Theorem 2.1 asserts that the sequence $(b_{p^i}(L))$ of the numbers of subrings of L of p-power index satisfies a strong regularity property. Indeed, a generating function of the form $\sum_{i=0}^{\infty} b_{p^i} Y^i$ is rational in the variable Y if and only if there is a finite linear recurrence relation on the coefficients b_{p^i}. The length of this recurrence relation is determined by the degree of the denominator polynomial. The numerator polynomial determines when the recurrence relation sets in; cf. [46, Theorem 4.1.1]. The numbers of finite index subalgebras of L_p are thus already determined by the numbers of subalgebras in some finite quotient of L_p.

A priori, Theorem 2.1 does not give us any information about further properties of the rational functions $W_p(Y)$. In particular, it does not tell us how the lengths of the relevant recurrence relations depend on the prime, or when they set in. In the examples above we observe that the denominators are all of the form $\prod_{i \in I}(1 - p^{a_i - b_i s})$ for non-negative integers a_i, b_i and a finite index set I. This, too, is a general phenomenon.

Theorem 2.2 ([10]). *For each $n \in \mathbb{N}$, there exists a finite index set I_n, and finitely many pairs $(a_i, b_i)_{i \in I_n}$ of natural numbers such that, if L is a ring of additive rank n, then for all primes p the denominator polynomial $Q_p(Y) \in \mathbb{Q}[Y]$ in Theorem 2.1 can be taken to divide $\prod_{i \in I_n}(1 - p^{a_i} Y^{b_i})$.*

Theorem 2.2 implies that the degrees in Y of the denominator polynomials $Q_p(Y)$ are bounded when L ranges over all rings of a given rank n. In particular, there is a uniform upper bound on the lengths of the recurrence relations satisfied by the sequences $(b_{p^i}(L))$ for fixed L as p ranges over the primes. It also shows that the coefficients of $Q_p(Y)$ are polynomials in p, so that the denominators of the Euler factors are polynomials in p and p^{-s}. The proof of Theorem 2.2 relies on non-constructive methods from model theory. No procedure is known to describe explicitly (even just a reasonably small superset of) the factors of the denominators of the local zeta functions of a given ring.

The numerators of the Euler factors have, in general, a far more complicated and interesting structure. In all of the examples we have encountered so far, the coefficients of the polynomials $P_p(Y)$, too, were – at least for almost all primes p – polynomials in p. It was known already to the authors of [22] that this is not a general feature. Their paper contains examples of zeta functions of nilpotent groups whose local factor at the prime p depends on how the rational prime p factorises in a number field. The right framework to explain this phenomenon, however, was not discovered until much later.

Theorem 2.3 ([17], Theorem 1.3). *Let L be a ring. There are smooth algebraic varieties V_t, $t \in \{1, \ldots, m\}$, defined over $\mathbb{Q}$, and rational functions $W_t(X, Y) \in \mathbb{Q}(X, Y)$ such that, for almost all primes p,*

$$\zeta_{L,p}(s) = \sum_{t=1}^{m} c_t(p) W_t(p, p^{-s}), \qquad (2.1)$$

where $c_t(p)$ denotes the number of $\mathbb{F}_p$-rational points of $\overline{V_t}$, the reduction modulo p of V_t.

Our formulation of Theorem 2.3 given here follows from the original formulation in [17] by the inclusion–exclusion principle. We will comment on the proof of this theorem at the end of Section 2.4. Note that the reduction modulo p of the varieties V_t only makes sense for all but finitely many primes p.

In general, the numbers of $\mathbb{F}_p$-rational points of the reduction modulo p of varieties defined over $\mathbb{Q}$ will not be polynomials in p, as the following example shows.

Example 2.4. Let E be the elliptic curve defined by the equation $y^2 = x^3 - x$. For an odd prime p, we denote by $c(p)$ the number of $\mathbb{F}_p$-rational points of $\overline{E}$, the reduction modulo p of E, i.e.

$$c(p) := \left| \left\{ (x, y) \in \mathbb{F}_p^2 \mid y^2 = x^3 - x \right\} \right|.$$

It is not hard to show that, if $p \equiv 3 \mod (4)$, then $c(p) = p$. If, however, $p \equiv 1 \mod (4)$, then $c(p) = p - (\pi + \overline{\pi})$, where π is the complex number satisfying $p = \pi\overline{\pi}$ and $\pi \equiv 1 \mod (2 + 2i)$; cf. [29, §18.4]. The discrepancy with the formula given in [29, §18.4, Theorem 5] comes from the fact that there $c(p)$ refers to the number of *projective* points of E, which includes also a point at infinity; cf. Example 2.15. This should also have been taken into account in [19, Example 1].

It is not clear *a priori* that varieties with such 'wild' non-polynomial reduction behaviour can occur in the description of zeta functions of rings given in (2.1). In [13, 14] du Sautoy gave an example of a class-2-nilpotent Lie ring (or, equivalently, class-2-nilpotent group) whose local zeta functions involve the cardinalities $c(p)$ associated to the elliptic curve in Example 2.4. In particular, he proved that the zeta function of this Lie ring is not 'finitely uniform'. We say that $\zeta_L(s)$ is *finitely uniform* if there are finitely many rational functions

$W_i(X, Y) \in \mathbb{Q}(X, Y)$, $i \in I$, a finite index set, such that for every prime p there exists an $i = i(p)$ such that $\zeta_{L,p}(s) = W_i(p, p^{-s})$. We say that $\zeta_L(s)$ is *uniform* if it is finitely uniform for $|I| = 1$ and *almost uniform* if there exists a rational function $W(X, Y)$ such that $\zeta_{L,p}(s) = W(p, p^{-s})$ for almost all p. We revisit du Sautoy's example in Section 3, where we look at the *ideal* zeta function of this particular Lie ring, counting ideals of finite index. For this variant, we give an explicit formula for the local zeta functions, illustrating Theorem 2.3 (or rather its analogue for ideal zeta functions of rings) in this particular case. At present, however, no explicit formula for the subgroup zeta function $\zeta_G(s)$ is known.

For future reference, we study in some detail an important family of varieties with 'uniform' reduction behaviour modulo p. It plays a key role in explicit formulae for zeta functions of rings.

2.2 Flag varieties and Coxeter groups

Let $n \in \mathbb{N}$, k be a field and let V denote an n-dimensional k-vector space. For each $i \in \{1, \ldots, n-1\}$, the set $G_{n,i}(k)$ of subspaces of V of dimension i can be given the structure of a smooth projective variety over k, called the ith Grassmannian variety of V. We define, for $i \in \{1, \ldots, n-1\}$, the polynomial

$$\binom{n}{i}_X := \prod_{j=0}^{i-1} (X^{n-j} - 1)/(X^{i-j} - 1) \in \mathbb{Z}[X].$$

It is not hard to prove that $|G_{n,i}(\mathbb{F}_q)| = \binom{n}{i}_q$. For example, the cardinality $|\mathbb{P}^{n-1}(\mathbb{F}_q)|$ of the $n-1$-dimensional projective space of lines in $\mathbb{F}_q^n$ is given by $\binom{n}{1}_q = (q^n - 1)/(q - 1) = 1 + q + \cdots + q^{n-1}$.

More generally, let $I = \{i_1, \ldots, i_l\}_<$ be a subset of $\{1, \ldots, n-1\}$. A *flag of type I* in V is a sequence $(V_i)_{i \in I}$ of subspaces of V satisfying

$$\{0\} \subsetneq V_{i_1} \subsetneq V_{i_2} \subsetneq \cdots \subsetneq V_{i_l} \subsetneq V$$

and, for all $i \in I$, $\dim_k(V_i) = i$. A flag is called *complete* if it is of type $I = \{1, \ldots, n-1\}$. For every $I \subseteq \{1, \ldots, n-1\}$, the set of flags of type I can be given the structure of a smooth projective variety over k. If $k = \mathbb{F}_q$, we obtain the variety of flags of type I in $\mathbb{F}_q^n$. We define the polynomial

$$\binom{n}{I}_X := \binom{n}{i_l}_X \binom{i_l}{i_{l-1}}_X \cdots \binom{i_2}{i_1}_X \in \mathbb{Z}[X]. \qquad (2.2)$$

One easily proves inductively that the number of flags of type I in $\mathbb{F}_q^n$ is given by the polynomial $\binom{n}{I}_q$, called *q-binomial coefficient* or *Gaussian polynomial*. For example, the number of complete flags in $\mathbb{F}_q^3$ is $(1 + q + q^2)(1 + q) = 1 + 2q + 2q^2 + q^3$.

For further applications, we shall need an expression for the polynomials $\binom{n}{I}_X$ in terms of Coxeter-group-theoretic notions.

Definition 2.5. Let S_n be the symmetric group of n letters with standard (Coxeter) generators $s_1, \ldots, s_{n-1}$. In cycle notation these are the transpositions $s_i = (i \; i+1)$. Let $w \in S_n$. The (*Coxeter*) *length* $\ell(w)$ is the length of a shortest word in the generators s_i representing w. The (*right*) *descent type* is the set

$$D_R(w) := \{i \in \{1, \ldots, n-1\} \mid w(i+1) < w(i)\}.$$

It can be shown that

$$D_R(w) = \{i \in \{1, \ldots, n-1\} \mid \ell(ws_i) < \ell(w)\}.$$

Example 2.6. The element $w = (1523) \in S_5$ has length 7. Indeed, a shortest word representing w is

$$s_2 s_3 s_1 s_4 s_3 s_2 s_1.$$

The descent type of w is $D_R(w) = \{1, 2, 4\}$.

Proposition 2.7. *Let q be a prime power. For all $I \subseteq \{1, \ldots, n-1\}$ we have*

$$\binom{n}{I}_q = \sum_{w \in S_n, \, D_R(w) \subseteq I} q^{\ell(w)}.$$

Proof. We first prove the proposition for $I = \{1, \ldots, n-1\}$. In this case, $\binom{n}{\{1,\ldots,n-1\}}_q$ is the number of complete flags $(V_i)_{i \in \{1,\ldots,n-1\}}$ in the finite vector space $\mathbb{F}_q^n$. These may be viewed as the cosets $\mathrm{GL}_n(\mathbb{F}_q)/B(\mathbb{F}_q)$, where B denotes the Borel subgroup consisting of upper-triangular matrices in GL_n. It is well known that the algebraic group GL_n satisfies a Bruhat decomposition

$$\mathrm{GL}_n = \bigcup_{w \in S_n} BwB$$

(where we identify permutations in S_n with permutation matrices in GL_n, acting from the left on unit column vectors, say). Therefore

$$\mathrm{GL}_n(\mathbb{F}_q)/B(\mathbb{F}_q) = \bigcup_{w \in S_n} B(\mathbb{F}_q)wB(\mathbb{F}_q)/B(\mathbb{F}_q).$$

The disjoint pieces $\Omega_w(\mathbb{F}_q) := B(\mathbb{F}_q)wB(\mathbb{F}_q)/B(\mathbb{F}_q)$, $w \in S_n$, are called Schubert cells; we refer the reader to [44, Chapter 8] and [40, Chapter 3] for further details on flag varieties and Schubert cells. It can be shown that each Schubert cell $\Omega_w(\mathbb{F}_q)$ is an affine space over $\mathbb{F}_q$ of $\mathbb{F}_q$-dimension $\ell(w)$. Indeed, a complete set of representatives of $B(\mathbb{F}_q)wB(\mathbb{F}_q)/B(\mathbb{F}_q)$, of size $q^{\ell(w)}$, is obtained in the following way: start with the permutation matrix corresponding to w. Substitute an arbitrary entry in $\mathbb{F}_q$ for each of the zeros of this matrix which is not positioned anywhere below or to the right of a 1. We conclude that

$$\binom{n}{\{1, \ldots, n-1\}}_q = |\mathrm{GL}_n(\mathbb{F}_q)/B(\mathbb{F}_q)| = \left| \bigcup_{w \in S_n} B(\mathbb{F}_q)wB(\mathbb{F}_q)/B(\mathbb{F}_q) \right|$$

$$= \sum_{w \in S_n} |\Omega_w(\mathbb{F}_q)| = \sum_{w \in S_n} q^{\dim_{\mathbb{F}_q}(\Omega_w(\mathbb{F}_q))} = \sum_{w \in S_n} q^{\ell(w)}.$$

This proves the proposition in the special case $I = \{1, \ldots, n-1\}$.

Example 2.8. Let $n = 5$. The Schubert cell Ω_w indexed by the element $w = (1523) \in S_5$ from Example 2.6 may be identified with the set of matrices of the form

$$\begin{pmatrix} * & * & 1 & 0 & 0 \\ * & * & 0 & * & 1 \\ * & 1 & 0 & 0 & 0 \\ * & 0 & 0 & 1 & 0 \\ 1 & 0 & 0 & 0 & 0 \end{pmatrix},$$

where $*$ may take any value in $\mathbb{F}_q$. Note that $\dim_{\mathbb{F}_q}(\Omega_w) = \ell(w) = 7$.

For general $I = \{i_1, \dots, i_l\}_< \subseteq \{1, \dots, n-1\}$, the quantity $\binom{n}{I}_q$ is the number of flags $(V_i)_{i \in I}$, $\dim_{\mathbb{F}_q}(V_i) = i$, in $\mathbb{F}_q^n$. These are in one-to-one correspondence with cosets $\mathrm{GL}_n(\mathbb{F}_q)/B_I(\mathbb{F}_q)$, where $B_I(\mathbb{F}_q)$ is the parabolic subgroup consisting of matrices of the form

$$\begin{pmatrix} \gamma_{i_1} & * & * & * \\ 0 & \gamma_{i_2 - i_1} & * & * \\ \vdots & \ddots & \ddots & \vdots \\ 0 & 0 & 0 & \gamma_{n-i_l} \end{pmatrix},$$

with $\gamma_i \in \mathrm{GL}_i(\mathbb{F}_q)$.

Among the Schubert cells $\Omega_w(\mathbb{F}_q)$ which are being identified by passing to cosets modulo $B_I(\mathbb{F}_q)$ there are unique ones of minimal dimension. It is not hard to see that these cells are exactly the cells indexed by elements $w \in S_n$ with $D_R(w) \subseteq I$, and that they constitute a set of representatives for the cosets $\mathrm{GL}_n(\mathbb{F}_q)/B_I(\mathbb{F}_q)$. We obtain

$$\binom{n}{I}_q = |\mathrm{GL}_n(\mathbb{F}_q)/B_I(\mathbb{F}_q)| = \left| \bigcup_{w \in S_n} B(\mathbb{F}_q) w B(\mathbb{F}_q)/B_I(\mathbb{F}_q) \right|$$

$$= \sum_{w \in S_n, D_R(w) \subseteq I} |\Omega_w(\mathbb{F}_q)| = \sum_{w \in S_n, D_R(w) \subseteq I} q^{\ell(w)}.$$

This proves Proposition 2.7 in general.   □

2.3   Counting with p-adic integrals

The idea to employ tools from the theory of p-adic integration to count subgroups and subrings is as old as the subject. It was first put to work in [22], and has been further developed ever since; see, for instance, [17] and [47]. All of the p-adic integrals used derive from Igusa's local zeta function, which we briefly review below. This will allow us to give a first proof of formula (1.4) for the zeta functions of abelian groups. We will also show how a formulation in terms of p-adic integrals enables us to express the local zeta functions of the Heisenberg Lie ring (cf. Example 1.3) in terms of the generating function associated to a polyhedral cone – or, equivalently, a system of linear homogeneous diophantine equations – which we can evaluate to establish formula (1.6). We return to generating functions of this kind in Section 2.4.

The p-adic integrals we consider are all variants of Igusa local zeta functions. Important references for this important class of p-adic integrals are [9, 28]. Given a polynomial $f = f(\mathbf{x}) \in \mathbb{Z}[x_1, \dots, x_n]$, Igusa's local zeta function associated to f is the p-adic integral

$$Z_f(s) := \int_{\mathbb{Z}_p^n} |f(\mathbf{x})|_p^s d\mu^{(n)}.$$

Here, $\mathbb{Z}_p$ are the p-adic integers, $\mu^{(n)}$ is the additive Haar measure on $\mathbb{Z}_p^n$, normalised so that $\mu^{(n)}\left(\mathbb{Z}_p^n\right) = 1$, $|\ |_p$ denotes the p-adic absolute value and s is a complex variable. For more background on the Haar measure on $\mathbb{Z}_p^n$, see, e.g., [15, Section 1.6].

The local zeta function associated to f is a powerful tool to understand the sequence (N_m), where, for $m \in \mathbb{N}_0$, we set

$$N_m := |\{\mathbf{x} \in (\mathbb{Z}/(p^m))^n \mid f(\mathbf{x}) = 0\}|.$$

These numbers may be encoded in a Poincaré series

$$P_f(t) := \sum_{m=0}^{\infty} p^{-nm} N_m t^m.$$

The Poincaré series $P_f(t)$ is related to the p-adic integral $Z_f(s)$ via the formula

$$P_f(p^{-s}) = \frac{1 - p^{-s} Z_f(s)}{1 - p^{-s}}. \tag{2.3}$$

Indeed, $p^{-nm} N_m$ is the measure of the set $\{\mathbf{x} \in \mathbb{Z}_p^n \mid v_p(f(\mathbf{x})) \geq m\}$ and thus

$$\mu^{(n)}\left(\{\mathbf{x} \in \mathbb{Z}_p^n \mid v_p(f(\mathbf{x})) = m\}\right) = p^{-nm} N_m - p^{-n(m+1)} N_{m+1}.$$

Thus

$$\begin{aligned}
Z_f(s) &= \sum_{m=0}^{\infty} \mu^{(n)}\left(\{\mathbf{x} \in \mathbb{Z}_p^n \mid v_p(f(\mathbf{x})) = m\}\right) p^{-sm} \\
&= \sum_{m=0}^{\infty} \left(p^{-nm} N_m - p^{-n(m+1)} N_{m+1}\right) p^{-sm} \\
&= P_f(p^{-s}) - p^s \left(P_f(p^{-s}) - 1\right) \\
&= (1 - p^s) P_f(p^{-s}) + p^s,
\end{aligned}$$

which is equivalent to (2.3). As an example we consider the integral

$$Z(s) := \int_{\mathbb{Z}_p} |x|_p^s d\mu^{(1)}$$

associated to $f(x) = x$. Observing that the associated Poincaré series equals

$$P(p^{-s}) = \sum_{m=0}^{\infty} (p^{-1-s})^m = \frac{1}{1 - p^{-1-s}} = \zeta_p(s+1)$$

we deduce that

$$Z(s) = \frac{1 - p^{-1}}{1 - p^{-1-s}} = (1 - p^{-1})\zeta_p(s+1). \tag{2.4}$$

We now explore how p-adic integrals may be used to count subgroups, by giving a first proof of formula (1.4). We may consider $\mathbb{Z}^n$ as a ring with trivial multiplication, so counting subgroups and counting subrings is the same thing in this case. Owing to the Euler product (1.2) it suffices to prove that, for each prime p, one has

$$\zeta_{\mathbb{Z}^n, p}(s) = \zeta_{\mathbb{Z}_p^n}(s) = \zeta_p(s)\zeta_p(s-1)\cdots\zeta_p(s-(n-1)). \tag{2.5}$$

The first equation is clear. For the second equation, assume that $\mathbb{Z}_p^n = \mathbb{Z}_p e_1 \oplus \ldots \oplus \mathbb{Z}_p e_n$ as a $\mathbb{Z}_p$-module, and set $\Gamma := \mathrm{GL}_n(\mathbb{Z}_p)$. Subgroups of $\mathbb{Z}_p^n$ of finite index may be identified with right Γ-cosets of $n \times n$-matrices over $\mathbb{Z}_p$ with non-zero determinant. Indeed, every such subgroup may be generated by n generators, whose coordinates with respect to the chosen basis may be encoded in the rows of an $n \times n$-matrix over $\mathbb{Z}_p$. Two such matrices M_1 and M_2 correspond to the same subgroup if and only if there is an element $\gamma \in \Gamma$ such that $M_1 = \gamma M_2$. In fact, one sees easily that these matrices may be chosen to lie in the set $\mathrm{Tr}_n(\mathbb{Z}_p)$ of upper-triangular matrices over $\mathbb{Z}_p$, so that subgroups H correspond to orbits $\mathcal{U}M =: \mathcal{M}(H)$, where $M \in \mathrm{Tr}_n(\mathbb{Z}_p) \cap \mathrm{GL}_n(\mathbb{Q}_p)$ and $\mathcal{U} := \Gamma \cap \mathrm{Tr}_n(\mathbb{Z}_p)$.

Now choose, for each $H \leq_f \mathbb{Z}_p^n$, a representative M_H in $\mathcal{U}M$, the $\mathcal{U}$-coset in $\mathrm{Tr}_n(\mathbb{Z}_p)$ corresponding to H. Notice that

$$\left| \mathbb{Z}_p^n : H \right| = |\det(M_H)|_p^{-1}, \tag{2.6}$$

and that

$$\mu(\mathcal{M}(H)) = (1 - p^{-1})^n \prod_{i=1}^{n} |(M_H)_{ii}|_p^i, \tag{2.7}$$

where μ denotes the additive Haar measure on $\mathrm{Tr}_n(\mathbb{Z}_p) \cong \mathbb{Z}_p^{\binom{n+1}{2}}$, normalised so that $\mu(\mathrm{Tr}_n(\mathbb{Z}_p)) = 1$. We thus obtain a partition

$$\mathrm{Tr}_n(\mathbb{Z}_p) = \bigcup_{H \leq_f \mathbb{Z}_p^n} \mathcal{M}(H) \cup \mathrm{Tr}_n^0(\mathbb{Z}_p), \tag{2.8}$$

where $\mathrm{Tr}_n^0(\mathbb{Z}_p)$ denotes the set of $n \times n$-upper-triangular matrices over $\mathbb{Z}_p$ with zero determinant. Note that $\mathrm{Tr}_n^0(\mathbb{Z}_p)$ has Haar measure zero. Using equations

(2.6), (2.7), (2.8) and (2.4) we compute

$$\sum_{H \leq \mathbb{Z}_p^n} |\mathbb{Z}_p^n : H|^{-s} = \sum_H |\det(M_H)|_p^s$$

$$= \sum_H \mu(\mathcal{M}(H))^{-1} \mu(\mathcal{M}(H)) \prod_{i=1}^n |(M_H)_{ii}|_p^s$$

$$= \sum_H (1 - p^{-1})^{-n} \prod_{i=1}^n |(M_H)_{ii}|_p^{-i} \int_{\mathcal{M}(H)} \prod_{i=1}^n |(M_H)_{ii}|_p^s d\mu$$

$$= (1 - p^{-1})^{-n} \int_{\mathrm{Tr}_n(\mathbb{Z}_p)} \prod_{i=1}^n |M_{ii}|_p^{s-i} d\mu$$

$$= (1 - p^{-1})^{-n} \int_{\mathbb{Z}_p^n} \prod_{i=1}^n |x_i|_p^{s-i} d\mu^{(n)}$$

$$= (1 - p^{-1})^{-n} \prod_{i=1}^n \int_{\mathbb{Z}_p} |x|_p^{s-i} d\mu^{(1)}$$

$$= \prod_{i=1}^n \zeta_p((s - i) + 1),$$

which proves (2.5).

Note that we computed each of the local factors of $\zeta_{\mathbb{Z}^n}(s)$ by expressing it as an integral over the affine space $\mathrm{Tr}_n(\mathbb{Z}_p)$ of upper-triangular matrices. The integrand in this integral is a simple function of the diagonal entries of the matrices. For arbitrary rings L, our above analysis carries through up to and including equation (2.7). In general, however, not every coset $\mathcal{U}M$ corresponds to a subring of L. We therefore need to describe conditions for such a coset to define a subring.

Let us return to Example 1.4 of the discrete Heisenberg group. Its associated Lie ring L has a $\mathbb{Z}$-basis (x, y, z), with relations $[x, y] = z$, $[x, z] = [y, z] = 0$. To compute the ring's local zeta function at the prime p, we need to count subalgebras in the $\mathbb{Z}_p$-algebra L_p. The rows of a matrix $M = (M_{ij}) \in \mathrm{Tr}_3(\mathbb{Z}_p) \cap \mathrm{GL}_3(\mathbb{Q}_p)$ encode the generators of a full additive sublattice Λ of $\mathbb{Z}_p^3$. To determine whether Λ is a subalgebra we need to check whether $\Lambda^2 \subseteq \Lambda$ or, equivalently, whether the lattice is closed under taking Lie brackets of its generators. In this case it is easy to see that the only condition we need to check is

$$[M_{11}x + M_{12}y + M_{13}z, M_{22}y + M_{23}z] \in \langle M_{33}z \rangle_{\mathbb{Z}_p}.$$

Using the commutator relation $[x, y] = z$ and the bilinearity of the Lie bracket $[\,,\,]$, we see that this condition is equivalent to

$$M_{33} \mid M_{11}M_{22}. \tag{2.9}$$

Note that this divisibility condition is equivalent to the inequality of p-adic valuations

$$v_p(M_{33}) \leq v_p(M_{11}) + v_p(M_{22}).$$

Setting $V_p := \{M \in \mathrm{Tr}_3(\mathbb{Z}_p) \mid M_{33} \mid M_{11}M_{22}\}$, we thus obtain

$$\zeta_{L_p}(s) = \sum_{H \leq L_p} |L_p : H|^{-s}$$

$$= (1 - p^{-1})^{-3} \int_{V_p} \prod_{i=1}^{3} |M_{ii}|_p^{s-i} d\mu^{(6)}$$

$$= (1 - p^{-1})^{-3} \int_{\{\mathbf{x} \in \mathbb{Z}_p^3 \mid x_3 \mid x_1 x_2\}} |x_1|_p^{s-1} |x_2|_p^{s-2} |x_3|_p^{s-3} d\mu^{(3)}$$

$$= \sum_{\{\mathbf{m} \in \mathbb{N}_0^3 \mid m_3 \leq m_1 + m_2\}} (p^{-s})^{m_1} (p^{1-s})^{m_2} (p^{2-s})^{m_3}.$$

It is not hard to compute this sum directly; see Proposition 2.12. It is instructive, however, to observe that it may be interpreted as a generating function associated to a system of linear homogeneous diophantine equations.

2.4   Linear homogeneous diophantine equations

Let $r, m \in \mathbb{N}_0$, let Φ be an $r \times m$ matrix over $\mathbb{Z}$ and consider the system of linear equations

$$\Phi \boldsymbol{\alpha} = \mathbf{0}, \tag{2.10}$$

where $\boldsymbol{\alpha}^t = (\alpha_1, \ldots, \alpha_m) \in \mathbb{Z}^m$ and $\mathbf{0} \in \mathbb{N}_0^r$. The non-negative integral solutions of (2.10) form a commutative monoid $\mathcal{E} := \{\boldsymbol{\alpha} \in \mathbb{N}_0^m \mid \Phi \boldsymbol{\alpha} = \mathbf{0}\}$ under addition. One approach to study this monoid is to investigate the generating function

$$E(\mathbf{X}) := E_\Phi(\mathbf{X}) := \sum_{\boldsymbol{\alpha} \in \mathcal{E}} \mathbf{X}^{\boldsymbol{\alpha}} \in \mathbb{Q}[[\mathbf{X}]],$$

where $\mathbf{X}^{\boldsymbol{\alpha}} = X_1^{\alpha_1} \cdots X_m^{\alpha_m}$ is a monomial in variables $X_1, \ldots, X_m$.

Generating functions of the form $E_\Phi(\mathbf{X})$ have been intensely studied by Stanley and others; cf. [46, Chapter 4.6], [45, Chapter I]. It can be proved, for example, that $E_\Phi(\mathbf{X})$ is always a rational function in the variables $X_1, \ldots, X_m$, with denominator of the form $\prod_{\beta \in CF(E)}(1 - \mathbf{X}^\beta)$, where β ranges over the finite set $CF(E)$ of *completely fundamental solutions* to Φ (compare with Theorem 2.2!). Here, a solution β to (2.10) is called *fundamental* if, whenever $\beta = \gamma + \delta$ for $\gamma, \delta \in \mathcal{E}$, $\gamma = \delta$ or $\delta = \beta$. A solution β to (2.10) is called *completely fundamental* if, whenever $n\beta = \gamma + \delta$ for $\gamma, \delta \in \mathcal{E}$ and $n \in \mathbb{N}$, then $\gamma = n_1 \beta$ for some $0 \leq n_1 \leq n$.

We further write $\overline{\mathcal{E}} := \{\boldsymbol{\alpha} \in \mathbb{N}^m \mid \Phi \boldsymbol{\alpha} = \mathbf{0}\}$ for the semigroup of positive integral solutions of (2.10) and consider the generating function

$$\overline{E}(\mathbf{X}) := \overline{E}_\Phi(\mathbf{X}) := \sum_{\boldsymbol{\alpha} \in \overline{\mathcal{E}}} \mathbf{X}^{\boldsymbol{\alpha}} \in \mathbb{Q}[[\mathbf{X}]].$$

The function $\overline{E}(\mathbf{X})$ is also rational in the variables $X_1, \ldots, X_m$. The following result of Stanley is of importance in applications to zeta functions of rings, in particular the proof of local functional equations. We denote by $1/\mathbf{X}$ the vector of inverted variables $(1/X_1, \ldots, 1/X_m)$. We also write $\mathscr{C}$ for the cone of non-negative real solutions to (2.10) and set $d := \dim_{\mathbb{R}}(\mathscr{C})$.

Theorem 2.9 ([46], Theorem 4.6.14). *Assume that $\overline{\mathcal{E}} \neq \varnothing$. Then*

$$\overline{E}(\mathbf{X}) = (-1)^d E(1/\mathbf{X}). \tag{2.11}$$

Example 2.10. Taking $r = 0$, or Φ the zero matrix, we obtain $\mathcal{E} = \mathbb{N}_0^m$, with completely fundamental solutions $\{(1, 0, \ldots, 0), \ldots, (0, \ldots, 0, 1)\}$, yielding

$$E(\mathbf{X}) = \sum_{\alpha \in \mathbb{N}_0^m} \mathbf{X}^\alpha = \prod_{i=1}^m \frac{1}{1 - X_i} \quad \text{and} \quad \overline{E}(\mathbf{X}) = \sum_{\alpha \in \mathbb{N}^m} \mathbf{X}^\alpha = \prod_{i=1}^m \frac{X_i}{1 - X_i}.$$

Example 2.11. Consider the matrix $\Phi = (1, 1, -1, -1)$. It can be shown that the (completely) fundamental solutions of the equation $\alpha_1 + \alpha_2 - \alpha_3 - \alpha_4 = 0$ are $(1, 0, 1, 0)$, $(1, 0, 0, 1)$, $(0, 1, 1, 0)$ and $(0, 1, 0, 1)$. Note that there is exactly one non-trivial relation between these solutions

$$(1, 0, 1, 0) + (0, 1, 0, 1) = (1, 0, 0, 1) + (0, 1, 1, 0) \, (= (1, 1, 1, 1)).$$

This can be used to show that

$$E_\Phi(X_1, X_2, X_3, X_4) = \frac{1 - X_1 X_2 X_3 X_4}{(1 - X_1 X_3)(1 - X_1 X_4)(1 - X_2 X_3)(1 - X_2 X_4)}.$$

See [45, I.11] for details.

Note that allowing inequalities rather than equalities in the system of equations (2.10) yields nothing more complicated. Indeed, an inequality may always be expressed in terms of an equality by introducing a slack variable. The generating functions enumerating integral points in rational polyhedral cones – i.e. intersections of finitely many rational half-spaces – may therefore be expressed in terms of generating functions associated to linear homogeneous diophantine equations. For example, we have $m_3 \leq m_1 + m_2$ if and only if there exists $m_4 \in \mathbb{N}_0$ such that $m_3 + m_4 = m_1 + m_2$ or, equivalently, $m_1 + m_2 - m_3 - m_4 = 0$. We obtain the generating function enumerating non-negative solutions of the inequality by taking the generating function associated to the equality by setting the variable corresponding to the slack variable to 1. From Example 2.11 we get, for instance

$$\sum_{\{\mathbf{m} \in \mathbb{N}_0^3 \mid m_3 \leq m_1 + m_2\}} X_1^{m_1} X_2^{m_2} X_3^{m_3} = E_\Phi(X_1, X_2, X_3, 1)$$

$$= \frac{1 - X_1 X_2 X_3}{(1 - X_1 X_3)(1 - X_1)(1 - X_2 X_3)(1 - X_2)}. \tag{2.12}$$

In Section 2.3 we showed how the local zeta functions of the Heisenberg Lie ring can be expressed in terms of the rational function given in (2.12). We summarise this result in:

Proposition 2.12 ([22], Proposition 8.1). *Let L be the Heisenberg Lie ring; cf. Example 1.3. Then, for all primes p, the local zeta function of L equals*

$$\zeta_{L_p}(s) = E_\Phi(p^{-s}, p^{1-s}, p^{2-s}, 1)$$

$$= \frac{1 - p^{3-3s}}{(1 - p^{-s})(1 - p^{1-s})(1 - p^{2-2s})(1 - p^{3-2s})}$$

$$= \zeta_p(s)\zeta_p(s-1)\zeta_p(2s-2)\zeta_p(2s-3)\zeta_p(3s-3)^{-1}.$$

We have thus expressed the local factors of the zeta function of the Heisenberg Lie ring in terms of the generating function associated with a linear homogeneous equation or, equivalently, a rational polyhedral cone. The feasibility of this approach is a direct consequence of the divisibility condition (2.9). In general, things are not that simple, as the following example shows.

Example 2.13. Let us reconsider the Lie ring $\mathfrak{sl}_2(\mathbb{Z})$ from Example 1.2. Fix a prime p. It is not hard – and a recommended exercise – to show that the coset $\mathcal{U}M$ of a matrix $M \in \mathrm{Tr}_3(\mathbb{Z}_p)$ encodes the coordinates of generators of a subring of $\mathfrak{sl}_2(\mathbb{Z}_p)$ if and only if

$$v_p(M_{22}) \le v_p(4M_{12}M_{23}),$$
$$v_p(M_{22}) \le v_p(4M_{12}M_{33}) \text{ and}$$
$$v_p(M_{22}M_{33}) \le v_p\left(M_{11}M_{22}^2 + 4M_{22}M_{13}M_{23} - 4M_{12}M_{23}^2\right);$$

cf. [20].

In general, the condition for a coset to define a subalgebra may be described by a finite number of inequalities in the p-adic values of polynomials in the matrix entries. If, as is the case for the Heisenberg Lie ring, these polynomials are monomials (cf. (2.9)), then the computation of the local zeta function reduces to the computation of the generating function of a rational polyhedral cone or, equivalently, a system of linear homogeneous diophantine equations.

Example 2.13 illustrates that the polynomials occuring need not be monomials. In this case, a resolution of singularities – a tool from algebraic geometry – may be used to remedy the situation. It allows for a partition of the domain of integration into pieces on which the integral may be expressed in terms of generating functions of polyhedral cones. The pieces are indexed by the $\mathbb{F}_p$-points of certain algebraic varieties defined over $\mathbb{F}_p$. These kinds of p-adic integrals, called 'cone integrals', were introduced in [17]. A more detailed introduction to cone integrals may be found in [19, Sections 4 and 5].

The description of local zeta functions of groups and rings in terms of cone integrals has far-reaching applications for the analysis of analytic properties of global zeta functions; cf. Section 2.7.

2.5   Local functional equations

The zeta functions of the rings we have presented so far as examples all share a remarkable property: their local factors generically exhibit a palindromic symmetry on inversion of the prime p. More precisely, almost all of the Euler factors satisfy a local functional equation of the form

$$\zeta_{L,p}(s)|_{p \to p^{-1}} = (-1)^a p^{b-cs} \zeta_{L,p}(s),$$

where a, b, c are integers which are independent of the prime p. In the present section we explain and give an outline of the proof of the following theorem:

Theorem 2.14 ([47], Theorem A). *Let L be a ring of additive torsion-free rank n. There are smooth projective varieties V_t, defined over $\mathbb{Q}$, and rational functions $W_t(X, Y) \in \mathbb{Q}(X, Y)$, $t \in \{1, \dots, m\}$, such that, for almost all primes p, the following hold:*

(1)

$$\zeta_{L,p}(s) = \sum_{t=1}^{m} b_t(p) W_t(p, p^{-s}), \tag{2.13}$$

where $b_t(p)$ denotes the number of $\mathbb{F}_p$-rational points of $\overline{V_t}$, the reduction modulo p of V_t.

(2) Setting $b_t(p^{-1}) := p^{-\dim(V_t)} b_t(p)$, the following functional equation holds:

$$\zeta_{L,p}(s)|_{p \to p^{-1}} = (-1)^n p^{\binom{n}{2}-ns} \zeta_{L,p}(s). \tag{2.14}$$

Note that the advance of Theorem 2.14 over Theorem 2.3 consists in the assertion (2.14). The notation '$p \to p^{-1}$' needs some explanation. If $b_t(p)$ is a polynomial in p, $b_t(p^{-1})$ is the rational number obtained by evaluating this polynomial at p^{-1}. This follows from the fact that the varieties $\overline{V_t}$ are smooth, projective varieties over finite fields. In general, the definition of $b_t(p^{-1})$ is motivated by properties of the numbers of $\mathbb{F}_p$-rational points of such varieties, which follow from the Weil conjectures. More precisely, let V be a smooth projective variety defined over the finite field $\mathbb{F}_p$. By deep properties of the Hasse–Weil zeta function associated to V, there are complex numbers α_{rj}, $0 \le r \le 2\dim(V)$, $1 \le j \le t_r$ for suitable non-negative integers t_r, such that the number $b_V(p)$ of $\mathbb{F}_p$-rational points of V can be written as

$$b_V(p) = \sum_{r=0}^{2\dim(V)} (-1)^r \sum_{j=1}^{t_r} \alpha_{rj}. \tag{2.15}$$

Note that the numbers t_r may well be zero; cf. the examples given in Section 2.2. Furthermore, for each $r \in \{0, \dots, 2\dim(V)\}$ the multisets

$$\{\alpha_{rj} | \, j \in \{1, \dots, t_{2\dim(V)-r}\}\} \quad \text{and} \quad \left\{ \frac{p^{\dim(V)}}{\alpha_{rj}} \Big| \, j \in \{1, \dots, t_r\} \right\}$$

coincide. Thus

$$b_V(p^{-1}) := p^{-\dim(V)} b_V(p) = \sum_{r=0}^{2\dim(V)} (-1)^r \sum_{j=1}^{t_r} \alpha_{rj}^{-1}$$

may be interpreted as the expression we obtain by inverting the terms α_{rj} in (2.15) even if they are not, in general, powers of the prime p.

Before we give an outline of the proof of Theorem 2.14, let us revisit Example 2.4.

Example 2.15. Let E denote the elliptic curve defined by the equation $y^2 = x^3 - x$; cf. Example 2.4. For an odd prime p denote by $b(p)$ the number of *projective* points of E over $\mathbb{F}_p$, i.e.

$$b(p) := |\{(x : y : z) \in \mathbb{P}^2(\mathbb{F}_p)|\, y^2 z = x^3 - xz^2\}|.$$

Clearly, $b(p) = c(p) + 1$, where $c(p)$ was defined in Example 2.4: we simply add the point $(0 : 1 : 0)$ 'at infinity'. The results quoted there imply that

$$b(p) = \begin{cases} 1 + p & \text{if } p \equiv 3 \mod (4) \text{ and} \\ 1 - (\pi + \overline{\pi}) + p & \text{otherwise,} \end{cases}$$

where $\pi\overline{\pi} = p$. Note that this last equation implies that $\pi^{-1} = \overline{\pi}/p$ and $\overline{\pi}^{-1} = \pi/p$, so that

$$b(p)|_{p \to p^{-1}} = p^{-1}b(p) = p^{-1}(p - (\overline{\pi} + \pi) + 1) = 1 - (\pi^{-1} + \overline{\pi}^{-1}) + p^{-1},$$

in the case that $p \equiv 1 \mod (4)$. If $p \equiv 3 \mod (4)$

$$b(p)|_{p \to p^{-1}} = p^{-1}b(p) = p^{-1}(1 + p) = p^{-1} + 1.$$

In the remainder of Section 2.5 we give an outline of the proof of Theorem 2.14. For details the reader is referred to [47, Sections 2 and 3]. The proof falls into two parts. The first is of a combinatorial and Coxeter-group-theoretic nature. It consists in proving the following general result about generating functions. We recall the definition of the polynomials $\binom{n}{I}_X \in \mathbb{Z}[X]$ given in (2.2).

Proposition 2.16. *Let* $n \in \mathbb{N}$ *and, let* $(W_I(p^{-s}))_{I \subseteq \{1,\dots,n-1\}}$ *be a family of functions in* p^{-s} *with the property that*

$$\forall I \subseteq \{1, \dots, n-1\} : \ W_I(p^{-s})|_{p \to p^{-1}} = (-1)^{|I|} \sum_{J \subseteq I} W_J(p^{-s}). \tag{2.16}$$

Then the function

$$W(p^{-s}) := \sum_{I \subseteq \{1,\dots,n-1\}} \binom{n}{I}_{p^{-1}} W_I(p^{-s}) \tag{2.17}$$

satisfies

$$W(p^{-s})|_{p \to p^{-1}} = (-1)^{n-1} p^{\binom{n}{2}} W(p^{-s}).$$

Remark 2.17. We do not need to specify the operation $p \to p^{-1}$ in (2.16) at this stage; the left-hand sides of these equations might well be defined in terms of the right-hand sides. Further, we do not assume the functions $W_I(p^{-s})$ to be rational in p^{-s}, or even in p and p^{-s}. In practice, we will apply Proposition 2.16 to families of rational functions $W_I(p^{-s})$, which are themselves of the form (2.13), and we define $p \to p^{-1}$ as in Theorem 2.14. What is understood, however, is that the inversion of the prime extends linearly to $W(p^{-s})$, and that $\binom{n}{I}_{p^{-1}}|_{p \to p^{-1}} = \binom{n}{I}_p$.

Proof of Proposition 2.16. We utilise the Coxeter-group-theoretic description of the numbers $\binom{n}{I}_p$ given in Proposition 2.7. Given $I \subseteq \{1, \ldots, n-1\}$ we write I^c for $\{1, \ldots, n-1\} \setminus I$. It is a standard fact (cf. [27, Section 1.8]) that there is a unique longest element $w_0 \in S_n$, namely the inversion such that, for all $w \in S_n$,

$$\ell(w) + \ell(ww_0) = \ell(w_0) = \binom{n}{2} \tag{2.18}$$

and

$$D_R(ww_0) = D_R(w)^c. \tag{2.19}$$

We also need the following lemma:

Lemma 2.18 ([51], Lemma 7). *Under the hypotheses of Proposition 2.16 we have, for all $I \subseteq \{1, \ldots, n-1\}$,*

$$\sum_{I \subseteq J} W_J(p^{-s})|_{p \to p^{-1}} = (-1)^{n-1} \sum_{I^c \subseteq J} W_J(p^{-s}).$$

Proof. We have

$$\sum_{I \subseteq J} W_J(p^{-s})|_{p \to p^{-1}} = \sum_{I \subseteq J} (-1)^{|J|} \sum_{T \subseteq J} W_T(p^{-s}) = \sum_{R \subseteq \{1, \ldots, n-1\}} c_R W_R(p^{-s}),$$

say, where

$$c_R = \sum_{R \cup I \subseteq J} (-1)^{|J|} = (-1)^{|R \cup I|} \sum_{T \subseteq (R \cup I)^c} (-1)^{|T|}$$

$$= (-1)^{|R \cup I|} (1-1)^{|R \cup I|^c} = \begin{cases} (-1)^{n-1} & \text{if } R \supseteq I^c, \\ 0 & \text{otherwise.} \end{cases}$$

This proves Lemma 2.18. □

Using (2.17), Proposition 2.7, (2.18), Lemma 2.18 and (2.19) we compute

$$
W(p^{-s})|_{p\to p^{-1}} = \sum_{I\subseteq\{1,\dots,n-1\}} \binom{n}{I}_p W_I(p^{-s})|_{p\to p^{-1}}
$$

$$
= \sum_{I\subseteq\{1,\dots,n-1\}} \left(\sum_{w\in S_n,\, D_R(w)\subseteq I} p^{\ell(w)} \right) W_I(p^{-s})|_{p\to p^{-1}}
$$

$$
= \sum_{w\in S_n} p^{\binom{n}{2}-\ell(ww_0)} \sum_{D_R(w)\subseteq I} W_I(p^{-s})|_{p\to p^{-1}}
$$

$$
= (-1)^{n-1}p^{\binom{n}{2}} \sum_{w\in S_n} p^{-\ell(ww_0)} \sum_{D_R(ww_0)\subseteq I} W_I(p^{-s})
$$

$$
= (-1)^{n-1}p^{\binom{n}{2}} \sum_{I\subseteq\{1,\dots,n-1\}} \left(\sum_{w\in S_n,\, D_R(ww_0)\subseteq I} p^{-\ell(ww_0)} \right) W_I(p^{-s})
$$

$$
= (-1)^{n-1}p^{\binom{n}{2}}W(p^{-s}).
$$

This proves Proposition 2.16.   □

Remark 2.19. The proof of Proposition 2.16 generalises readily to general finite Coxeter groups. In [33] this is used to study enumeration problems in geometric algebra.

The second part of the proof of Theorem 2.14 consists in proving that the local zeta function $\zeta_{L,p}(s)$ of a ring L of additive torsion-free rank n may be written as

$$
(1 - p^{-ns})^{-1}W(p^{-s}),
$$

where $W(p^{-s})$ is of the form (2.17) for suitable (rational) functions $W_I(p^{-s})$, satisfying the hypotheses (2.16) of Proposition 2.16. This requires both algebro-geometric and combinatorial methods which are similar to but markedly different from the ones used to study cone integrals. It may be instructive to see this done in a familiar special case first.

Example 2.20. We will see in Example 2.21 that the local zeta functions of the abelian group $\mathbb{Z}^n$ may be written as

$$
\zeta_{\mathbb{Z}^n,p}(s) = \frac{1}{1-X_n} \sum_{I\subseteq\{1,\dots,n-1\}} \binom{n}{I}_{p^{-1}} \prod_{\iota\in I} \frac{X_\iota}{1-X_\iota},
$$

where, for $i \in \{1,\dots,n\}$, $X_i := p^{i(n-i)-is}$. One checks immediately that the functions

$$
W_I(p^{-s}) := \prod_{\iota\in I} \frac{X_\iota}{1-X_\iota} \tag{2.20}
$$

satisfy (2.16). Indeed, the operation $p \to p^{-1}$ simply amounts to an inversion of the 'variables' X_i, as they are monomials in p and p^{-s}, and

$$\frac{X^{-1}}{1 - X^{-1}} = -\left(1 + \frac{X}{1 - X}\right).$$

More conceptually, the validity of the equations (2.16) may be regarded as a consequence of Theorem 2.9 in the special case studied in Example 2.10, as we may view $W_I(p^{-s})$ as obtained from the rational generating function in variables X_i, counting positive integral solutions of an (empty) set of linear homogeneous diophantine equations in $|I|$ variables, where the variables X_i are substituted by monomials in p and p^{-s}. A variation of this basic idea is crucial for the proof of Theorem 2.14.

Given a prime p, the zeta function $\zeta_{L,p}(s)$ enumerates full additive sublattices of the n-dimensional $\mathbb{Z}_p$-algebra $L_p \subset L_p \otimes_{\mathbb{Z}_p} \mathbb{Q}_p$ which are subalgebras, i.e. which are closed under multiplication. It is easy to see that, given any full lattice Λ, there is a unique lattice Λ_0 in the *homothety class* $[\Lambda] := \mathbb{Q}_p^* \Lambda$ of Λ such that the subalgebras contained in $[\Lambda]$ are exactly the multiples $p^m \Lambda_0$, $m \in \mathbb{N}_0$. Indeed, given any sublattice Λ of L_p, and $e \in \mathbb{Z}$, clearly $(p^e \Lambda)^2 \subseteq p^e \Lambda$ if and only if $p^e \Lambda^2 \subseteq \Lambda$. Clearly, there always exists an integer e satisfying this condition. Let $e_0 := \min\{e \in \mathbb{Z} | p^e \Lambda \subseteq L_p \text{ and } p^e \Lambda^2 \subseteq \Lambda\}$, and set $\Lambda_0 := p^{e_0} \Lambda$. Evidently, Λ_0 only depends on the homothety class of Λ. We thus have

$$\zeta_{L_p}(s) = (1 - p^{-ns})^{-1} \sum_{[\Lambda]} |L_p : \Lambda_0|^{-s}.$$

We set

$$W(p^{-s}) := \sum_{[\Lambda]} |L_p : \Lambda_0|^{-s}.$$

It remains to show that $W(p^{-s})$ is of the form (2.17), with rational functions $W_I(p^{-s})$ to which Proposition 2.16 is applicable. We achieve this by first partitioning the set of homothety classes of lattices into finitely many parts, indexed by the subsets I of $\{1, \ldots, n-1\}$, reflecting (aspects of) their elementary divisor types. On each of these parts, we describe the indices $|L_p : \Lambda_0|$ in terms of algebraic congruences, and then encode the numbers of solutions to these congruences in terms of a suitable p-adic integral $W_I(p^{-s})$ so that the family $(W_I(p^{-s}))_{I \subseteq \{1, \ldots, n-1\}}$ satisfies the 'inversion properties' (2.16). The proof of the latter requires sophisticated methods from algebraic geometry which we can only sketch here.

The reader will note the analogy with the proof of equation (2.3) which also proceeded by expressing the numbers of certain congruences in terms of the Haar measure of suitable sets.

We recall from Section 2.3 that $\mathbb{Z}_p$-submodules of L_p of finite index are in one-to-one correspondence with cosets ΓM, where $\Gamma = \mathrm{GL}_n(\mathbb{Z}_p)$ and $M \in \mathrm{Tr}_n(\mathbb{Z}_p) \cap \mathrm{GL}_n(\mathbb{Q}_p)$; the rows of M encode coordinates of generators of Λ with respect to a fixed basis $(e_1, \ldots, e_n)$ for L_p as a $\mathbb{Z}_p$-module. For $r \in \{1, \ldots, n\}$, let C_r denote the matrix of the linear map given by right-multiplication with the basis element e_r with respect to this basis. It is not hard to show (cf. the proof of [17, Theorem 5.5]) that the lattice corresponding to the coset ΓM is a sub*algebra* if and only if

$$\forall i,j \in \{1,\ldots,n\} : M_i \sum_{r=1}^{n} C_r m_{jr} \in \langle M_k| \ k \in \{1,\ldots,n\}\rangle_{\mathbb{Z}_p}, \qquad (2.21)$$

where M_i denotes the ith row of M. This condition is easy to check if M is diagonal; in this case, condition (2.21) is satisfied if, for all $k \in \{1,\ldots,n\}$, the kth entries of all the vectors on the left-hand side are divisible by M_{kk}, the kth diagonal entry of M. In general, however, the coset ΓM will not contain a diagonal element. One way around this is to choose a different basis for L_p. Indeed, by the elementary divisor theorem, the coset ΓM contains an element of the form $D\alpha^{-1}$, where $\alpha \in \Gamma$ and

$$D = D(I,\mathbf{r}_0) = p^{r_0} \operatorname{diag}(\underbrace{\underbrace{p^{\sum_{\iota \in I} r_\iota},\ldots,p^{\sum_{\iota \in I} r_\iota}}_{i_1},\ldots,p^{r_{i_l}},\ldots,p^{r_{i_l}},1,\ldots,1)}_{i_l}$$

for a set $I = \{i_1,\ldots,i_l\}_< \subseteq \{1,\ldots,n-1\}$ and a vector $(r_0,r_{i_1},\ldots,r_{i_l}) =: \mathbf{r}_0 \in \mathbb{N}_0 \times \mathbb{N}^l$, both depending only on ΓM. Setting $\mathbf{r} := (r_{i_1},\ldots,r_{i_l})$, we say that the homothety class $[\Lambda]$ of Λ is of type $(I,\mathbf{r})$ or sometimes, by abuse of notation, of type I. We write $\nu([\Lambda]) = (I,\mathbf{r})$ or $\nu([\Lambda]) = I$, respectively. The matrix α is only unique up to right-multiplication by an element of

$$\Gamma_{I,\mathbf{r}} := \left\{ \left(\begin{array}{c|c|c|c|c} \gamma_{i_1} & * & \cdots & * & * \\ \hline p^{r_{i_1}}* & \gamma_{i_2-i_1} & \ddots & \vdots & \vdots \\ \hline p^{r_{i_1}+r_{i_2}}* & p^{r_{i_2}}* & \ddots & * & \vdots \\ \hline \vdots & \vdots & \ddots & \gamma_{i_l-i_{l-1}} & * \\ \hline p^{r_{i_1}+\cdots+r_{i_l}}* & p^{r_{i_2}+\cdots+r_{i_l}}* & \cdots & p^{r_{i_l}}* & \gamma_{n-i_l} \end{array} \right) \right\},$$

where $\gamma_i \in \Gamma_i := \mathrm{GL}_i(\mathbb{Z}_p)$, and $*$ stands for an arbitrary matrix of the appropriate size with entries in $\mathbb{Z}_p$. As a corollary, we deduce a formula for the number of lattices of given type $(I,\mathbf{r})$

$$|\{[\Lambda]| \ \nu([\Lambda]) = (I,\mathbf{r})\}| = |\Gamma : \Gamma_{I,\mathbf{r}}| = \mu(\Gamma)/\mu(\Gamma_{I,\mathbf{r}}) = \binom{n}{I}_{p^{-1}} p^{\sum_{\iota \in I} r_\iota \iota(n-\iota)}.$$

$$(2.22)$$

Here, μ denotes the Haar measure on the group Γ normalised so that $\mu(\Gamma) = (1-p^{-1})\cdots(1-p^{-n})$. It is a crucial observation that this Haar measure coincides with the restriction of the additive Haar measure $\mu^{(n^2)}$ on $\mathrm{Mat}_n(\mathbb{Z}_p) \cong \mathbb{Z}_p^{n^2}$, normalised so that $\mu^{(n^2)}(\mathrm{Mat}_n(\mathbb{Z}_p)) = 1$.

We denote by λ_{ij}^k the structure constants of L with respect to the chosen basis, i.e. $e_i e_j = \sum_{k=1}^{n} \lambda_{ij}^k e_k$, and define linear forms $\mathrm{L}_{ij}(\mathbf{y}) := \sum_{k=1}^{n} \lambda_{ij}^k y_k \in$

$\mathbb{Z}[\mathbf{y}]$. We consider the $n \times n$-matrix

$$\mathcal{R}(\mathbf{y}) = (\mathrm{L}_{ij}(\mathbf{y})) \in \mathrm{Mat}_n(\mathbb{Z}[\mathbf{y}]).$$

Writing $\alpha[i]$ for the ith column of the matrix α and setting $\mathcal{R}_{(i)}(\alpha) := \alpha^{-1}\mathcal{R}(\alpha[i])(\alpha^{-1})^t$ we may now, after right-multiplication by α, rewrite the sub-algebra condition (2.21) equivalently to

$$\forall i \in \{1, \ldots, n\} : \ D\mathcal{R}_{(i)}(\alpha)D \equiv 0 \mod (D_{ii}), \tag{2.23}$$

as a quick calculation shows. Considering these matrix congruences modulo a common modulus, equation (2.23) is equivalent to

$$\forall i, r, s \in \{1, \ldots, n\} :$$
$$(\mathcal{R}_{(i)}(\alpha))_{rs} \, p^{r_0 + \sum_{s \leq \iota \in I} r_\iota + \sum_{r \leq \iota \in I} r_\iota + \sum_{i > \iota \in I} r_\iota} \equiv 0 \mod (p^{\sum_{\iota \in I} r_\iota}). \tag{2.24}$$

Setting, for $(i, r, s) \in \{1, \ldots, n\}^3$

$$v_{irs}(\alpha) := \min \left\{ v_p \left((\mathcal{R}_{(\iota)}(\alpha))_{\rho\sigma} \right) | \iota \leq i, \rho \geq r, \sigma \geq s \right\}$$

and

$$m([\Lambda]) :=$$

$$\min \left\{ \sum_{\iota \in I} r_\iota, \ \sum_{s \leq \iota \in I} r_\iota + \sum_{r \leq \iota \in I} r_\iota + \sum_{i > \iota \in I} r_\iota + v_{irs}(\alpha) | \, (i, r, s) \in \{1, \ldots, n\}^3 \right\},$$
$$\tag{2.25}$$

condition (2.24) may in turn be reformulated as

$$r_0 \geq \sum_{\iota \in I} r_\iota - \mu([\Lambda]).$$

We observe that the definition (2.25) of the quantity $m([\Lambda])$ is in terms which are linear in the $(r_\iota)_{\iota \in I}$ and the terms $v_{irs}(\alpha)$, which only depend on α. Moreover, by construction the $v_{irs}(\alpha)$ only depend on the coset αB, where $B \subseteq \mathrm{GL}_n(\mathbb{Z}_p)$ is the Borel subgroup of upper-triangular matrices. This is the purpose of using inequalities rather than equalities in the definition of the terms $v_{irs}(\alpha)$. As in the proof of the identity (2.3), we may now express the numbers of lattice classes $[\Lambda]$ of given type $\nu([\Lambda])$ and invariant $m([\Lambda])$ in terms of the Haar measure of the set on which the integrand of a certain p-adic integral is constant. More precisely, we set, for $(i, r, s) \in \{1, \ldots, n\}^3$

$$\mathbf{f}_{irs}(\mathbf{y}) := \{(\mathcal{R}_{(\iota)}(\mathbf{y}))_{\rho\sigma} | \ \iota \leq i, \rho \geq r, \sigma \geq s\}$$

and

$$Z_I((s_\iota)_{\iota \in I}, s_n) :=$$

$$\int_{p\mathbb{Z}_p^I \times \Gamma} \prod_{\iota \in I} |x_\iota|_p^{s_\iota} \left\| \left\{ \prod_{\iota \in I} x_\iota \right\} \cup \bigcup_{(i,r,s)} \left(\prod_{\iota \in I} x_\iota^{\delta_{\iota \geq r} + \delta_{\iota \geq s} + \delta_{\iota < i}} \right) \mathbf{f}_{irs}(\mathbf{y}) \right\|_p^{s_n} d\mathbf{x}_I d\mathbf{y}.$$
$$\tag{2.26}$$

Here we extend the p-adic absolute value to a set $\mathcal{S}$ of p-adic numbers by setting $\|\mathcal{S}\|_p := \max\{|s|_p|\ s \in \mathcal{S}\}$, denote by $d\mathbf{x}_I = dx_{i_1} \cdots dx_{i_l}$ the Haar measure on $p\mathbb{Z}_p^l$, and write δ_P for the 'Kronecker delta' which is equal to 1 if the property P holds and equal to 0 otherwise.

This p-adic integral has been set up so that, for each $I \subseteq \{1, \ldots, n-1\}$, we have

$$\sum_{\nu([\Lambda])=I} |L_p : \Lambda_0|^{-s} = \binom{n}{I}_{p-1} W_I(p^{-s}),$$

say, where

$$W_I(p^{-s}) := \frac{Z_I((s(\iota+n) - \iota(n-\iota) - 1)_{\iota \in I}, -sn)}{(1-p^{-1})^l \mu(\Gamma)},$$

so that

$$W(p^{-s}) = \sum_{I \subseteq \{1,\ldots,n-1\}} \binom{n}{I}_{p-1} W_I(p^{-s}).$$

We need to establish that the functions $W_I(p^{-s})$ satisfy the inversion property (2.16). Let us first confirm this in the abelian case.

Example 2.21. If L_p is abelian, i.e. if the multiplication on L_p is trivial, all the sets of polynomials $\mathbf{f}_{irs}$ are equal to $\{0\}$, so (2.26) takes the form

$$Z_I((s_\iota)_{\iota \in I}, s) = \int_{(p\mathbb{Z}_p)^l \times \Gamma} \prod_{\iota \in I} |x_\iota|_p^{s_\iota + s_n} \, d\mathbf{x}_I d\mathbf{y}$$

$$= \mu(\Gamma) \prod_{\iota \in I} \int_{p\mathbb{Z}_p} |x_\iota|_p^{s_\iota + s_n} \, dx_\iota = \mu(\Gamma)(1-p^{-1})^l \prod_{\iota \in I} \frac{p^{-1-s_\iota - s_n}}{1 - p^{-1-s_\iota - s_n}}$$

and thus

$$W_I(p^{-s}) = \prod_{\iota \in I} \frac{p^{\iota(n-\iota)-s\iota}}{1 - p^{\iota(n-\iota)-s\iota}},$$

in accordance with (2.20). Of course, we could have deduced this immediately from (2.22), avoiding any reference to p-adic integrals.

We note that in the formula (2.26), the variables $\mathbf{x}$ enter monomially. If the same was true for the variables $\mathbf{y}$, the inversion properties (2.16) would follow from the following proposition, generalising a result of Stanley:

Proposition 2.22 ([47], Proposition 2.1). *Let $s, t \in \mathbb{N}_0$ and, for $\sigma \in \{1, \ldots, s\}$, $\tau \in \{1, \ldots, t\}$, let $L_{\sigma\tau}(\mathbf{n})$ be $\mathbb{Z}$-linear forms in the variables $n_1, \ldots, n_r$. Let $X_1, \ldots, X_r, Y_1, \ldots, Y_s$ be independent variables and set*

$$Z^\circ(\mathbf{X}, \mathbf{Y}) := \sum_{\mathbf{n} \in \mathbb{N}^r} \prod_{\rho=1}^r X_\rho^{n_\rho} \prod_{\sigma=1}^s Y_\sigma^{\min\{L_{\sigma\tau}(\mathbf{n})|\tau \in \{1,\ldots,t\}\}},$$

$$Z(\mathbf{X}, \mathbf{Y}) := \sum_{\mathbf{n} \in \mathbb{N}_0^r} \prod_{\rho=1}^r X_\rho^{n_\rho} \prod_{\sigma=1}^s Y_\sigma^{\min\{L_{\sigma\tau}(\mathbf{n})|\tau \in \{1,\ldots,t\}\}}.$$

Then

$$Z^\circ(\mathbf{X}^{-1}, \mathbf{Y}^{-1}) = (-1)^r Z(\mathbf{X}, \mathbf{Y}).$$

For $t \leq 1$, this follows immediately from Theorem 2.9, as $Z^\circ(\mathbf{X}, \mathbf{Y})$ and $Z(\mathbf{X}, \mathbf{Y})$ may be interpreted in terms of the generating functions $\overline{E}(\mathbf{X})$ and $E(\mathbf{X})$, respectively, associated to the empty set of equations in r variables. The general case follows from an adaptation of the proof of [46, Proposition 4.16.14].

In general, the inversion properties (2.16) can be proved by making the integral (2.26) 'locally monomial' in the variables $\mathbf{y}$. This is achieved by applying a 'principalisation of ideals', a tool from algebraic geometry. More precisely, we apply the following deep result to the ideal $\prod_{(i,r,s) \in \{1,\ldots,n\}^3} (\mathbf{f}_{irs}(\mathbf{y}))$, defining a subvariety of the homogeneous space $X = \mathrm{GL}_n / B$.

Theorem 2.23 ([53], Theorem 1.0.1). *Let $\mathfrak{I}$ be a sheaf of ideals on a smooth algebraic variety X. There exists a principalisation (Y, h) of $\mathfrak{I}$, i.e. a sequence*

$$X = X_0 \xleftarrow{h_1} X_1 \longleftarrow \cdots \xleftarrow{h_\iota} X_\iota \longleftarrow \cdots \xleftarrow{h_r} X_r = Y$$

of blow-ups $h_\iota : X_\iota \to X_{\iota-1}$ of smooth centres $C_{\iota-1} \subset X_{\iota-1}$ such that:

(a) *The exceptional divisor E_ι of the induced morphism $h^\iota = h_\iota \circ \cdots \circ h_1 : X_\iota \to X$ has only simple normal crossings and C_ι has simple normal crossings with E_ι.*

(b) *Setting $h := h_r \circ \cdots \circ h_1$, the total transform $h^*(\mathfrak{I})$ is the ideal of a simple normal crossing divisor $\widetilde{E}$. If the subscheme determined by $\mathfrak{I}$ has no components of codimension one, then $\widetilde{E}$ is an $\mathbb{N}$-linear combination of the irreducible components of the divisor E_r.*

The existence of a principalisation lies as deep as Hironaka's celebrated resolution of singularities in characteristic 0 [24]; see [53] for details.

2.6 A class of examples: 3-dimensional p-adic anti-symmetric algebras

Constructing an explicit principalisation for a given family of ideals $(\mathbf{f}_{irs}(\mathbf{y}))_{irs}$ is in general very difficult. In the special case that L_p is an anti-symmetric (not necessarily nilpotent or Lie) $\mathbb{Z}_p$-algebra of dimension 3, however, the approach of Theorem 2.14 leads to an explicit, unified expression for the zeta function of L_p.

Theorem 2.24 ([32], Theorem 1). *Let L be a 3-dimensional anti-symmetric $\mathbb{Z}_p$-algebra. Then there is a ternary quadratic form $f = f(\mathbf{x}) \in \mathbb{Z}_p[x_1, x_2, x_3]$, unique up to equivalence, such that, for $i \geq 0$*

$$\zeta_{p^i L}(s) = \zeta_{\mathbb{Z}_p^3}(s) - Z_f(s-2)\zeta_p(2s-2)\zeta_p(s-2)p^{(2-s)(i+1)}(1 - p^{-1})^{-1},$$

where $Z_f(s)$ is Igusa's local zeta function associated to f.

The form $f(\mathbf{x})$ in Theorem 2.24 may be defined explicitly in terms of the structure constants of L_p with respect to a chosen basis. Different bases give rise to equivalent forms; see [32] for details.

This result yields, in particular, a uniform expression for the zeta functions of all 3-dimensional $\mathbb{Z}_p$-Lie algebras we have seen so far (and others, e.g. [31]). For example, the forms $f(\mathbf{x})$ for the abelian algebra $\mathbb{Z}_p^3$, the Heisenberg Lie algebra and the 'simple' Lie algebra $\mathfrak{sl}_2(\mathbb{Z}_p)$ are 0, x_3^2 and $x_3^2 - 4x_1 x_2$, respectively.

Using the setup of Section 2.5, the key to proving Theorem 2.24 is the observation that only the functions $W_I(p^{-s})$ with $1 \in I$ differ from the 'abelian' functions (2.20). Indeed, if $1 \notin I$, the conditions (2.24) hold for all $r_0 \in \mathbb{N}_0$. If $1 \in I$, then they hold if and only if

$$p^{r_0}(\mathcal{R}_{(1)}(\alpha))_{23} \equiv 0 \mod (p^{r_1}) \tag{2.27}$$

and a quick calculation shows that, for $\alpha = (\alpha_{ij}) \in \Gamma_3$, one has

$$\det(\alpha)(\mathcal{R}_{(1)}(\alpha))_{23} = L_{23}(\alpha[1])\alpha_{11} - L_{13}(\alpha[1])\alpha_{21} + L_{12}(\alpha[1])\alpha_{31}.$$

Setting
$$f(\mathbf{x}) := L_{23}(\mathbf{x})x_1 - L_{13}(\mathbf{x})x_2 + L_{12}(\mathbf{x})x_3$$

we see that (2.27) holds if and only if

$$r_0 \geq r_1 - v_p(f(\alpha[1])).$$

The computation of the integral (2.26) is thus no harder than the computation of the Igusa zeta function associated to the quadratic polynomial $f(\mathbf{x})$.

We note that Theorem 2.24 also yields a complete description of the possible poles of zeta functions of 3-dimensional $\mathbb{Z}_p$-Lie algebras, as the poles of Igusa's local zeta function of quadratic forms are well understood; cf. [32, Corollary 1.2]. In higher dimensions, such a description is entirely elusive. Theorem 2.24 also shows explicitly the relationship between $\zeta_L(s)$ and $\zeta_{pL}(s)$ if L is of dimension 3. No such formula is known in higher dimensions.

2.7 Global zeta functions of groups and rings

Let G be a group with polynomial subgroup growth. As noted in the introduction, the degree of polynomial subgroup growth of G is encoded in an analytic invariant of the group's zeta function $\zeta_G(s)$, namely its abscissa of convergence $\alpha = \alpha(G)$. The zeta function's analytic properties hold, however, the key to more information about the subgroup growth of G. The following is a deep result.

Theorem 2.25 ([17], Theorem 1.1). *Let G be a $\mathfrak{T}$-group:*

1. *The abscissa of convergence α is a rational number, and $\zeta_G(s)$ can be meromorphically continued to $\mathrm{Re}(s) > \alpha - \delta$ for some $\delta > 0$. The continued function is holomorphic on the line $\mathrm{Re}(s) = \alpha$ except for a pole at $s = \alpha$.*

2. *Let $b+1$ denote the multiplicity of the pole of $\zeta_G(s)$ at $s = \alpha$. There exists a constant $c \in \mathbb{R}$ such that*

$$s_m(G) \sim c \cdot m^\alpha (\log m)^b \quad as \ m \to \infty.$$

The proof of Theorem 2.25 given in [17] proceeds via an analysis of the local 'cone integrals' mentioned above.

Whilst it is a remarkable fact that global zeta functions of nilpotent groups always allow for some meromorphic continuation beyond their abscissa of convergence, it is not the case that they may all be continued to the whole complex plane, as is the case for abelian groups or the Heisenberg group. In fact, numerous groups have been found for which there are natural boundaries for meromorphic continuation; cf. [21, Chapter 7]. Surprisingly little is known about the abscissa of convergence α and the pole order $b + 1$ in general.

3 Variations on a theme

The theme of counting subobjects of finite index in a nilpotent group or a ring may be varied in several interesting ways.

3.1 Normal subgroups and ideals

One of the forerunners of the very concept of the zeta function of a group is the Dedekind zeta function of a number field, one of the most classical objects in algebraic number theory. Given a number field k with ring of integers $\mathcal{O}$, the Dedekind zeta function of k is defined as the Dirichlet series

$$\zeta_k(s) := \sum_{\mathfrak{a} \lhd_f \mathcal{O}} |\mathcal{O} : \mathfrak{a}|^{-s},$$

where the sum ranges over the ideals of finite index in $\mathcal{O}$ and s is a complex variable. Owing to our understanding of the ideal structure in the Dedekind ring $\mathcal{O}$, we have good control of arithmetic and analytic properties of this important function. In particular, we know that it has an Euler product indexed by the prime ideals in $\mathcal{O}$, that it allows for an analytic continuation to the whole complex plane and has a simple pole at $s = 1$. Its residue at this pole encodes important arithmetic information about the number field k, given by the class number formula; see, for instance, [7].

The ideal zeta function of a general ring L is defined as the Dirichlet series

$$\zeta_L^{\lhd}(s) := \sum_{m=1}^{\infty} b_m^{\lhd} m^{-s} = \sum_{H \lhd_f L} |L : H|^{-s},$$

where $b_m^{\lhd} = b_m^{\lhd}(L)$ is the number of ideals in L of index m. Similarly, if G is a group of polynomial *normal* subgroup growth, it is of interest to study the normal zeta function of G, which is defined as

$$\zeta_G^{\triangleleft}(s) := \sum_{m=1}^{\infty} a_m^{\triangleleft} m^{-s} = \sum_{H \triangleleft_f G} |G : H|^{-s},$$

where $a_m^{\triangleleft} = a_m^{\triangleleft}(G)$ denotes the number of normal subgroups of G of index m.

Both the ideal zeta function of a ring L and the normal zeta function of a nilpotent group G satisfy a Euler product decomposition

$$\zeta_L^{\triangleleft}(s) = \prod_{p \text{ prime}} \zeta_{L,p}^{\triangleleft}(s), \qquad \zeta_G^{\triangleleft}(s) = \prod_{p \text{ prime}} \zeta_{G,p}^{\triangleleft}(s)$$

into local factors enumerating subobjects of p-power index. The study of normal subgroup growth of G can also be linearised using the Lie ring introduced in Section 1.3. By [22, Section 4] we have, for almost all primes p,

$$\zeta_{G,p}^{\triangleleft}(s) = \zeta_{L,p}^{\triangleleft}(s), \tag{3.1}$$

where $L = L(G)$ is the nilpotent Lie ring associated to G.

Example 3.1. Let G be the discrete Heisenberg group from Example 1.4. It can be shown that

$$\zeta_G^{\triangleleft}(s) = \zeta_L^{\triangleleft}(s) = \zeta(s)\zeta(s-1)\zeta(3s-2).$$

See Exercise 5.3 and [22, Section 8]. Note that the equation (3.1) holds for all primes p.

In many ways, the theory of ideal zeta functions of nilpotent groups is similar to the theory of their subgroup zeta functions. In particular, the local factors $\zeta_{G,p}^{\triangleleft}(s)$ are also rational in p^{-s}, and analogues of Theorems 2.2, 2.3 and 2.25 hold. The first explicitly computed example of a non-uniform zeta function is the normal zeta function of a class-2-nilpotent group.

Example 3.2. In [13] du Sautoy showed that both the subgroup and the normal subgroup zeta function of the following class-2-nilpotent group are not finitely uniform. He defined

$$\mathcal{R}(\mathbf{y}) = \begin{pmatrix} 0 & R(\mathbf{y}) \\ -R(\mathbf{y})^{\mathrm{t}} & 0 \end{pmatrix} \in \mathrm{Mat}_6(\mathbb{Z}[y_1, y_2, y_3]) \text{ with } R(\mathbf{y}) = \begin{pmatrix} y_3 & y_1 & y_2 \\ y_1 & y_3 & 0 \\ y_2 & 0 & y_1 \end{pmatrix}$$

and set

$$G := \langle x_1, \ldots, x_6, y_1, y_2, y_3 | \ \forall i, j : [x_i, x_j] = \mathcal{R}(\mathbf{y})_{ij}, \quad \text{all other } [,] \text{ trivial} \rangle,$$

using additive notation for words in the abelian subgroup of G generated by y_1, y_2, y_3. Notice that the polynomial $\det(R(\mathbf{y})) = y_1 y_3^2 - y_1^3 - y_2^2 y_3$ defines the projective elliptic curve E considered in Example 2.15. In [48, p. 1031] it is shown that, for $p \neq 2$, one has

$$\zeta^{\triangleleft}_{G,p}(s) = \zeta_{\mathbb{Z}^6_p}(s)(W_1(p, p^{-s}) + b(p)W_2(p, p^{-s})),$$

with $b(p)$ defined in Example 2.15 and

$$W_1(X, Y) = \frac{1 + X^6Y^7 + X^7Y^7 + X^{12}Y^8 + X^{13}Y^8 + X^{19}Y^{15}}{(1 - X^{18}Y^9)(1 - X^{14}Y^8)(1 - X^8Y^7)}$$

$$W_2(X, Y) = \frac{(1 - Y^2)X^6Y^5(1 + X^{13}Y^8)}{(1 - X^{18}Y^9)(1 - X^{14}Y^8)(1 - X^8Y^7)(1 - X^7Y^5)}.$$

Using the identity $b(p)|_{p \to p^{-1}} = p^{-1}b(p)$, the functional equation

$$\zeta^{\triangleleft}_{G,p}(s)|_{p \to p^{-1}} = -p^{36-15s}\zeta^{\triangleleft}_{G,p}(s) \tag{3.2}$$

follows immediately. The local subgroup zeta functions $\zeta_{G,p}(s)$ have not been calculated explicitly; cf. Problem 4.4.

The methods used to perform the calculations in Example 3.2 rely on the fact that the square root of the determinant of the matrix of relations $\mathcal{R}(\mathbf{y})$ defines a smooth hypersurface in the projective space over the centre of the group. Together with the algebro-geometric fact that every smooth plane curve defined over $\mathbb{Q}$ may be defined by the determinant of a suitable matrix of linear forms, one can, in this way, force any such curve to take on the role played by the elliptic curve in Example 3.2 in the normal zeta function of a class-2-nilpotent group. We refer to [49] for details.

Equation (3.2) is a special case of an analogue of Theorem 2.14 for normal zeta functions of class-2-nilpotent Lie rings [47, Theorem C]. To what extent this symmetry phenomenon extends to normal zeta functions of other (Lie) rings is largely mysterious. Examples due to Woodward show that this may or may not hold in Lie rings of higher nilpotency classes, and in certain soluble Lie rings; cf. [21]. We refer to the related problems and conjectures in Section 4.1

3.2 Representations

By counting subgroups of a group, we count essentially permutation representations of the group. A natural variant on the theme of counting subgroups of a given group therefore consists in enumerating the group's finite-dimensional irreducible complex representations. In this context, too, the concept of a zeta function is a helpful enumeration tool if the group has – at least up to some equivalence relation – only finitely many irreducible complex representations of each finite dimension, and if these numbers grow at most polynomially. We call an (abstract or profinite) group G *rigid* if, for every $m \in \mathbb{N}$, the number $r_m(G)$ of isomorphism classes of (continuous, if G is profinite,) irreducible complex representations of G of dimension n is finite. We say that a rigid group G has *polynomial representation growth* (PRG) if, for each $m \in \mathbb{N}$, the number of representations of G of dimension at most m is bounded above by a polynomial in m. As in the case of counting subgroups, we define the *representation zeta function* as the Dirichlet generating function

$$\zeta_G^{\mathrm{irr}}(s) := \sum_{m=1}^{\infty} r_m(G) m^{-s} = \sum_{\rho} (\dim(\rho))^{-s},$$

where ρ ranges over the isomorphism classes of finite-dimensional irreducible complex representations of G. It defines a convergent function on the complex half-plane determined by the infimum of the numbers $\alpha \in \mathbb{R}$ such that $\sum_{i \leq m} r_i(G) = O(1 + m^{\alpha})$.

No general characterisation of rigid or PRG groups is known. In the current section we will concentrate on results regarding three classes of groups: finitely generated torsion-free nilpotent (or $\mathfrak{T}$-)groups, semisimple arithmetic groups and compact p-adic analytic groups.

As we will see in Section 3.2, $\mathfrak{T}$-groups are 'rigid up to twisting with 1-dimensional representations'. The growth of the numbers of the ensuing equivalence classes, called 'twist-isoclasses', is polynomial, and the associated representation zeta functions satisfy Euler product decompositions, indexed by the primes, analogous to the context of counting subgroups. The Kirillov orbit method offers a suitable 'linearisation' of the problem of counting twist-isoclasses of representations of p-power dimension, and we may once again use our arsenal of tools from p-adic integration to study the Euler factors at least for almost all primes.

In Section 3.2 we briefly review results that show that the representation zeta functions of certain arithmetic groups also satisfy an Euler product decomposition, indexed by all places of a number field. The factors indexed by non-archimedean places are zeta functions associated to compact p-adic analytic groups. As we shall see, these are also rational functions, albeit not solely in the parameter p^{-s}.

$\mathfrak{T}$-groups

A non-trivial $\mathfrak{T}$-group has infinitely many 1-dimensional irreducible representations: it has infinite abelianisation, and the group of 1-dimensional representations of $\mathbb{Z}^n$, i.e. of homomorphisms of $\mathbb{Z}^n$ to $\mathbb{C}^*$, is isomorphic to $(\mathbb{C}^*)^n$. Tensoring with 1-dimensional representations will thus yield an infinitude of m-dimensional representations for every m for which such representations exist. Fortunately, this turns out to be all that needs fixing. More precisely, given a $\mathfrak{T}$-group G and $m \in \mathbb{N}$, we denote by $R_m(G)$ the set of m-dimensional irreducible complex representations of G and s is a complex variable. Given $\sigma_1, \sigma_2 \in R_m(G)$, we say that σ_1 and σ_2 are *twist–equivalent* if there exists a 1-dimensional representation $\chi \in R_1(G)$ such that $\sigma_1 = \chi \otimes \sigma_2$. The classes of this equivalence relation are called *twist-isoclasses*. The set $R_m(G)$ has the structure of a quasi–affine complex algebraic variety whose geometry was analysed by Lubotzky and Magid. They proved in [36, Theorem 6.6] that there is a finite quotient $G(m)$ of G such that every m-dimensional irreducible representation of G is twist-equivalent to one that factors through $G(m)$. In particular, the number $c_m = c_m(G)$ of twist-isoclasses of irreducible m-dimensional representations is finite. In [26], the representation zeta function of G is defined by

$$\zeta_G^{\mathrm{irr}}(s) := \sum_{m=1}^{\infty} c_m m^{-s}.$$

Furthermore, the function $m \mapsto c_m$ is multiplicative. Indeed, this follows from Lubotzky and Magid's result together with the group-theoretic fact that the finite nilpotent groups $G(m)$ are the direct products of their Sylow p-subgroups and the basic representation-theoretic fact that the irreducible representations of direct products of finite groups are exactly the tensor products of irreducible representations of their factors; cf. [8, (10.33)]. Thus

$$\zeta_G^{\mathrm{irr}}(s) = \prod_{p \text{ prime}} \zeta_{G,p}^{\mathrm{irr}}(s), \quad \text{where} \quad \zeta_{G,p}^{\mathrm{irr}}(s) := \sum_{i=0}^{\infty} c_{p^i} p^{-is}.$$

As in the case of saturable pro-p-groups (see Section 10.2 in Chapter I), there is a close connection between representations of $\mathfrak{T}$-groups and co-adjoint orbits. This generalisation of Kirillov's orbit method to the realm of $\mathfrak{T}$-groups is due to Howe. In [25] he shows that (twist-isoclasses of) irreducible representations in a $\mathfrak{T}$-group G are parametrised by co-adjoint orbits of certain (additive) characters on the associated Lie ring $L = L(G)$. More precisely, we write $\widehat{L}$ for the group $\mathrm{Hom}_{\mathbb{Z}}(L, \mathbb{C}^*)$, and Ad^* for the co-adjoint action of G on $\widehat{L}$. We denote by L' the Lie subring of L corresponding to the group's derived group G'. We say that a character $\psi \in \widehat{L}$ is rational on L' if its restriction to L' is a torsion element, i.e. if $\psi(nL') \equiv 1$ for some $n \in \mathbb{N}$. The smallest such n is called the period of ψ. Howe's principal result states that a character's co-adjoint orbit is finite if and only if the character is rational on L', and that finite Ad^*-orbits $\Omega \subseteq \widehat{L}$ of characters of odd period are in one-to-one correspondence with (twist-isoclasses of) finite-dimensional representations U_Ω of G of dimension $|\Omega|^{1/2}$; see [47, Section 3.4] for details.

To effectively enumerate twist-isoclasses of finite-dimensional representations of G we thus have to deal with two problems: given a character $\psi \in \widehat{L}$ of finite period, we firstly need to determine the size of its co-adjoint orbit. Secondly, to control over-counting, we have to determine the size of the co-adjoint orbit of the restriction of ψ to L'. From now on, we will restrict ourselves to the case that the nilpotency class of G is 2. In this case, the latter task is trivial as the co-adjoint action on the restriction of characters to L', which is central, is trivial.

As in the case of saturable pro-p groups, we associate with a character $\psi \in \widehat{L}$ the bi-additive antisymmetric map

$$b_\psi : L \times L \to \mathbb{C}^*, \quad (x, y) \mapsto \psi([x, y]).$$

Note that b_ψ only depends on the restriction of ψ to L'. We define

$$\text{Rad}_\psi := \text{Rad}(b_\psi) = \{x \in L| \ \forall y \in L : b_\psi(x,y) = 1\}.$$

One can show that if ψ is rational on L' (so its co-adjoint orbit is finite by Howe's result) and $|L : \text{Rad}|$ is coprime to finitely many 'bad primes', depending only on G, then Rad_ψ is the Lie ring corresponding to the stabiliser subgroup $\text{Stab}_G(\psi)$ of ψ under the co-adjoint action. Then, by the orbit stabiliser theorem, the index $|L : \text{Rad}_\psi|$ equals the size of the co-adjoint orbit of ψ. The Kirillov correspondence now implies that the representation associated to the orbit of ψ has degree $|\Omega|^{-1/2} = |G : \text{Stab}_G(\psi)|^{-1/2} = |L : \text{Rad}_\psi|^{-1/2}$.

Recall that, for a class-2-nilpotent group, finite co-adjoint orbits are parametrised by rational characters on L' of finite period. For a prime p and $N \in \mathbb{N}_0$, we write Ψ_N for the set of characters on $\widehat{L'}$ of period p^N. By Howe's results we have:

Theorem 3.3 ([47], Corollary 3.1). *Let G be a class-2-nilpotent $\mathfrak{T}$-group. Then, for almost all odd primes p*

$$\zeta_{G,p}^{\text{irr}}(s) = \sum_{N \in \mathbb{N}_0, \ \psi \in \Psi_N} |L : \text{Rad}_\psi|^{-s/2}. \qquad (3.3)$$

Assume that $\text{rk}(G/G') = d$ and $\text{rk}(G') = d'$, say, and let p be a prime for which (3.3) holds. To compute the right-hand side of this equation effectively, we identify Ψ_N with $W_{p,N} := (\mathbb{Z}/(p^N))^{d'} \setminus p(\mathbb{Z}/(p^N))^{d'}$ as additive groups, and let $\mathcal{R}(\mathbf{y}) \in \text{Mat}(d, \mathbb{Z}[y_1, \ldots, y_{d'}])$ be the matrix of linear forms encoding the commutator structure of G, i.e. $\mathcal{R}(\mathbf{y})_{ij} = \sum_{k=1}^{d'} \lambda_{ij}^k y_k$ if G is generated by $e_1, \ldots, e_d$ subject to the relations $[e_i, e_j] = \sum_{k=1}^{d'} \lambda_{ij}^k f_k$, say, where $G/G' = \langle e_1 G', \ldots, e_d G' \rangle$ and $G' = \langle f_1, \ldots, f_{d'} \rangle$.

A simple computation shows that if $\psi \in \Psi_N$ corresponds to $\ell \in W_{p,N}$, then the index of Rad_ψ in L equals the index of the system of linear congruences

$$\mathcal{R}(\ell)\mathbf{x} \equiv 0 \mod (p^N),$$

where $\mathbf{x} \in \mathbb{Z}_p^d$. This index can be easily computed from the elementary divisors of the matrix $\mathcal{R}(\ell)$. Recall that $\mathcal{R}(\ell)$ is said to have elementary divisor type $\mathbf{m} = (m_1, \ldots, m_d) \in \{0, \ldots, N\}^d$ – written $\nu(\mathcal{R}(\ell)) = \mathbf{m}$ – if there are matrices $\beta, \gamma \in \text{GL}_d(\mathbb{Z}/p^N)$ such that

$$\beta\mathcal{R}(\ell)\gamma \equiv \begin{pmatrix} p^{m_1} & & \\ & \ddots & \\ & & p^{m_d} \end{pmatrix}$$

and $0 \leq m_1 \leq \cdots \leq m_d \leq N$. Given $N \in \mathbb{N}_0$ and $\mathbf{m} \in \mathbb{N}_0^d$ we set

$$\mathcal{N}_{N,\mathbf{m}} := |\{\ell \in W_{p,N}| \ \nu(\mathcal{R}(\ell)) = \mathbf{m}\}|.$$

It is now easy to see that

$$\zeta_{G,p}^{\mathrm{irr}}(s) = \sum_{N\in\mathbb{N}_0,\ \mathbf{m}\in\mathbb{N}_0^d} \mathcal{N}_{N,\mathbf{m}} p^{-\sum_{i=1}^d (N-m_i)s/2}. \tag{3.4}$$

This 'Poincaré series' may, in analogy to the prototype equation (2.3), be expressed in terms of a p-adic integral. The integrand of this – in general quite complicated – integral is defined in terms of the minors of the matrix $\mathcal{R}(\mathbf{y})$. This approach yields immediately the rationality of almost all of the local representation zeta functions of $\mathcal{T}$-groups. Rationality was first established, for all primes p, by model-theoretic means in [26].

The general case of $\mathcal{T}$-groups of arbitrary nilpotency class is complicated by having to account for over-counting when we run over the characters of L'. This can also be formulated in terms of elementary divisors of matrices of forms; cf. [47, Section 2.2]. We illustrate the computations in class 2 outlined above with a familiar example.

Example 3.4. Let G be the discrete Heisenberg group from Example 1.4. Here $d = 2$ and $d' = 1$. For all primes p and $N \in \mathbb{N}_0$, we have $W_{p,N} = (\mathbb{Z}/(p^N))^*$. The commutator matrix $\mathcal{R}(\mathbf{y})$ is given by

$$\mathcal{R}(y) = \begin{pmatrix} & y \\ -y & \end{pmatrix}$$

and therefore

$$\mathcal{N}_{N,\mathbf{m}} = \begin{cases} 1 & \text{if } N = 0, \\ (1-p^{-1})p^N & \text{if } N \in \mathbb{N} \text{ and } m_1 = m_2 = 0, \\ 0 & \text{otherwise.} \end{cases}$$

Thus, for all primes p

$$\zeta_{G,p}^{\mathrm{irr}}(s) = \sum_{N\in\mathbb{N}_0,\mathbf{m}\in\mathbb{N}_0^2} \mathcal{N}_{N,\mathbf{m}} p^{-Ns+(m_1+m_2)s/2} \tag{3.5}$$

$$= 1 + \sum_{N\in\mathbb{N}} (1-p^{-1})p^{(1-s)N}$$

$$= (1-p^{-s})/(1-p^{1-s}),$$

or, equivalently

$$\zeta_G^{\mathrm{irr}}(s) = \sum_{m=1}^{\infty} \varphi(m)m^{-s} = \zeta(s-1)\zeta(s)^{-1},$$

where φ denotes the Euler totient function. This was first proved in [41, Theorem 5] by a direct calculation of the twist-isoclasses.

Notice that the local factors of the representation zeta function of the Heisenberg group all satisfy the functional equation

$$\zeta_{G,p}^{\mathrm{irr}}(s)|_{p\to p^{-1}} = p\,\zeta_{G,p}^{\mathrm{irr}}(s).$$

This generalises in the following way:

Theorem 3.5 ([47], Theorem D). *Let G be a $\mathcal{T}$-group with derived group G' of Hirsch length d'. Then, for almost all primes p*

$$\zeta_{G,p}^{\mathrm{irr}}(s)|_{p\to p^{-1}} = p^{d'}\,\zeta_{G,p}^{\mathrm{irr}}(s). \tag{3.6}$$

Semisimple arithmetic groups

Let k be a number field with ring of integers $\mathcal{O}$, and S be a finite set of places of k, including all the archimedean ones. We denote by $\mathcal{O}_S$ the S-integers in k. Let $\mathbf{G}$ be a semisimple, simply connected and connected algebraic group $\mathbf{G}$ defined over k, together with a fixed embedding into GL_N for some $N \in \mathbb{N}$. By an arithmetic group we mean, in the following, a group G which is commensurable to $\mathbf{G}(\mathcal{O}_S)$. Recall that G is said to have the congruence subgroup property (CSP) if every finite index subgroup of G is a congruence subgroup; see Section 3 of Chapter II. Recall further that, if G is rigid, the representation zeta function $\zeta_G^{\mathrm{irr}}(s)$ of G has finite abscissa of convergence if and only if G has polynomial representation growth (PRG).

Theorem 3.6 ([38], Theorems 1.2 and 1.3). *Let G be an arithmetic group. Then G has PRG if and only it has the CSP.*

Assume from now on that G is an arithmetic group with the CSP. For a non-archimedean place v of k we write $\mathcal{O}_v$ for the completion of $\mathcal{O}$ at v.

Proposition 3.7 ([34], Proposition 4.6). *There is a subgroup G_0 of G of finite index in G such that*

$$\zeta_{G_0}^{\mathrm{irr}}(s) = \zeta_{\mathbf{G}(\mathbb{C})}^{\mathrm{irr}}(s)^{|S_\infty|} \prod_{v\notin S} \zeta_{L_v}^{\mathrm{irr}}(s), \tag{3.7}$$

where S_∞ denotes the set of archimedean valuations of k, L_v is an open subgroup of $\mathbf{G}(\mathcal{O}_v)$ and $\zeta_{\mathbf{G}(\mathbb{C})}^{\mathrm{irr}}(s)$, respectively $\zeta_{L_v}^{\mathrm{irr}}(s)$, enumerates irreducible rational, respectively continuous, representations of $\mathbf{G}(\mathbb{C})$, respectively L_v.

The fact that we need to pass to a finite index subgroup in Proposition 3.7 is insubstantial if we are mainly interested in the representation zeta function's abscissa of convergence. Indeed, we have the following:

Lemma 3.8 ([34], Corollary 4.5). *If G is a PRG group and $G_0 \leq G$ a finite index subgroup, then the abscissae of convergence of the zeta functions $\zeta_G^{\mathrm{irr}}(s)$ and $\zeta_{G_0}^{\mathrm{irr}}(s)$ coincide.*

Example 3.9. Let $G = \mathrm{SL}_n(\mathbb{Z})$. It is well known that $\mathrm{SL}_n(\mathbb{Z})$ satisfies the CSP if and only if $n \geq 3$. In this case, Proposition 3.7 yields that

$$\zeta_{\mathrm{SL}_n(\mathbb{Z})}^{\mathrm{irr}}(s) = \zeta_{\mathrm{SL}_n(\mathbb{C})}^{\mathrm{irr}}(s) \prod_{p \text{ prime}} \zeta_{\mathrm{SL}_n(\mathbb{Z}_p)}^{\mathrm{irr}}(s).$$

Already at first glance the Euler product (3.7) differs from the Euler factorisations we have encountered before by the presence of factors 'at infinity'. The Euler factor $\zeta_{\mathbf{G}(\mathbb{C})}^{\mathrm{irr}}(s)$ is, however, comparatively well understood. For instance, we have:

Theorem 3.10 ([34], Theorem 5.1). *The abscissa of convergence of $\zeta_{\mathbf{G}(\mathbb{C})}^{\mathrm{irr}}(s)$ is equal to ρ/κ, where $\rho = \mathrm{rk}(\mathbf{G})$ and $\kappa = |\Phi^+|$ is the number of positive roots.*

The proof of Theorem 3.10 is based the fact that the rational representations of these groups are combinatorially parametrised by their highest weights; see [34, Section 5] for details.

Example 3.11. The group $\mathrm{SL}_2(\mathbb{C})$ has a unique irreducible rational representation of each finite dimension. Thus

$$\zeta_{\mathrm{SL}_2(\mathbb{C})}^{\mathrm{irr}} = \sum_{m=1}^{\infty} m^{-s} = \zeta(s).$$

The abscissa of convergence of the Riemann zeta function is $1 = 1/1 = \rho/\kappa$.

Regarding the abscissa of convergence of the global representation zeta function of an arithmetic group, the following is known:

Theorem 3.12 ([2], Theorem 1.2). *Let G be an arithmetic group with the CSP. Then the abscissa of convergence of $\zeta_G^{\mathrm{irr}}(s)$ is a rational number.*

The proof of this deep result uses sophisticated tools from algebraic geometry, model theory and the representation theory of finite groups of Lie type. We only remark that whilst its conclusion is analogous to one of the conclusions of Theorem 2.25, its proof requires substantially different methods.

Compact p-adic analytic groups

The groups L_v in Proposition 3.7 are compact p-adic analytic groups. Let, more generally, G be a finitely generated profinite group. It is well known that the numbers $r_m(G)$ of isomorphism classes of continuous irreducible n-dimensional complex representations of G are all finite if and only if G is FAb, i.e. if and only if every open subgroup of G has finite abelianisation; cf. Section 10.1 in Chapter I.

Theorem 3.13 ([30], Theorem 1.1). *Let G be a compact FAb p-adic analytic group with $p > 2$. Then there are natural numbers $n_1, \ldots, n_k$ and rational functions $f_1(Y), \ldots, f_k(Y) \in \mathbb{Q}[Y]$ such that*

$$\zeta_G^{\mathrm{irr}}(s) = \sum_{i=1}^{k} n_i^{-s} f_i(p^{-s}).$$

This deep result takes a more complicated form than the rationality results for Euler factors we have met before. It should not surprise us, however, that the representation zeta function of a p-adic analytic group is not, in general, a rational function just in p^{-s}: whereas the continuous representations of a pro-p group clearly all have dimension a power of p (as they factor over finite index normal subgroups of the group), a p-adic analytic group is only virtually pro-p, i.e. it has a pro-p subgroup of finite index. The natural numbers $n_1, \ldots, n_k$ in Theorem 3.13 can be interpreted as the dimensions of the representations of the quotient of G by a normal, finite index pro-p subgroup.

As the work on representation zeta functions for $\mathcal{T}$-groups sketched in Section 3.2, the proof of Theorem 3.13 is based on a Kirillov orbit method for compact p-adic analytic groups.

Explicit examples of representation zeta functions of compact p-adic groups are difficult to compute. In [30], Jaikin computes $\zeta_{\mathrm{SL}_2(R)}^{\mathrm{irr}}(s)$, where R is any complete discrete valuation ring with odd residue field cardinality. Note, however, that the group $\mathrm{SL}_2(\mathbb{Z})$, for instance, does not have the CSP. The Euler product $\zeta_{\mathrm{SL}_2(\mathbb{C})}(s) \prod_{p \text{ prime}} \zeta_{\mathrm{SL}_2(\mathbb{Z})}(s)$ does hence *not* represent the representation zeta function of $\mathrm{SL}_2(\mathbb{Z})$, but only accounts for the 'congruence representations' of $\mathrm{SL}_2(\mathbb{Z})$, i.e. those which factor over a congruence quotient of $\mathrm{SL}_2(\mathbb{Z})$.

The paper [5] develops a p-adic formalism for representation zeta functions of certain p-adic analytic pro-p groups, using the Kirillov orbit method as a key tool; see also [3]. For FAb pro-p groups it provides a formula of the form (2.13) for the representation zeta functions. By arguments similar to the ones in the proof of Theorem 2.14 we deduce functional equations akin to (3.6) for representation zeta functions of pro-p groups in globally defined families; see also Theorem 10.3 in Chapter I. In noteworthy contrast to subgroup growth zeta functions, the results in [5] show that the representation zeta functions of the relevant groups satisfy a strong 'uniformity' property under ring extensions. This phenomenon implies monotonicity results for the relevant abscissae of convergence. It is illustrated by the explicit formulae for the representation zeta functions of groups of the form $\mathrm{SL}_3(\mathcal{O}_v)$ given in [4], valid in case the residue characteristic is not 3. The formulae are uniform in q, the residue field cardinality of $\mathcal{O}_v$, and exemplify the functional equations. Using delicate arguments involving Clifford theory, the paper [5] also determines the abscissa of convergence of representation zeta functions of arithmetic groups in algebraic groups of type A_2 defined over number fields.

3.3 Further variations

Nilpotent groups

Besides the zeta functions counting all subgroups, normal subgroups and representations of a $\mathfrak{T}$-group G, people have studied the zeta functions enumerating subgroups of G which are isomorphic to G, the 'pro-isomorphic' zeta functions enumerating subgroups whose profinite completion is isomorphic to the profinite completion of G and the zeta functions enumerating subgroups up to conjugacy; cf. [22, 18, 6, 47]. The last two types of zeta functions satisfy Euler product decompositions into Euler factors which are rational in p^{-s}.

Compact p-adic analytic groups

Let G be a compact p-adic analytic group. Recall that such a group is virtually pro-p. In [11] du Sautoy proved that the 'local' zeta function

$$\zeta_{G,p}(s) = \sum_{n=0}^{\infty} a_{p^n}(G) p^{-ns}$$

of G is rational in p^{-s}. He also proved that the 'global' zeta function $\zeta_G(s)$ counting all finite-index subgroups is rational in $p^{-s}, n_1^{-s}, \dots, n_k^{-s}$ for natural numbers $n_1, \dots, n_k$ (analogous to Theorem 3.13), and established similar results for zeta functions counting normal subgroups, r-generator subgroups and subgroups up to conjugacy in compact p-adic analytic groups. We refer to [39, Chapter 16] for details. In [16] du Sautoy showed the rationality of certain generating functions enumerating the class numbers of (i.e. the total numbers of conjugacy classes in) families of finite groups associated to compact p-adic analytic groups.

Finite p-groups

The methods used to study the subgroup growth of nilpotent or p-adic analytic groups have found applications in the enumeration of finite p-groups, notably Higman's PORC conjecture; see [12] and Section 4.2 of Chapter I. Given a prime p and natural numbers c, d and n, let $f(n, p, c, d)$ denote the number of (isomorphism classes of) d-generator p-groups of order p^n and nilpotency class at most c. We define the Dirichlet generating function

$$\zeta_{c,d,p}(s) := \sum_{n=0}^{\infty} f(n, p, c, d) p^{-ns}.$$

In [12] du Sautoy proved that these generating series are rational in the parameter p^{-s}. The connection with zeta functions of nilpotent groups is the following.

Every p-group on d generators of nilpotency class c is a quotient of the free class-c-nilpotent group $F := F_{c,d}$ on d generators by a normal subgroup N of p-power index. Two such groups may, of course, define isomorphic quotients. In [12] he manages to control this ambiguity by considering open normal subgroups in the pro-p completion $(\widehat{F_{c,d}})_p$. It is a fact that two such subgroups M and N define isomorphic quotients if and only if there exists an automorphism φ of $(\widehat{F_{c,d}})_p$ such that $\varphi(M) = N$; cf. [39, Proposition 16.4.1]. To count isomorphism classes of the finite quotients it therefore suffices to enumerate orbits of this automorphism group acting on the lattice of open normal subgroups of $(\widehat{F_{c,d}})_p$. Computing the local normal zeta functions $\zeta^\triangleleft_{F_{c,d},p}(s)$ may thus be seen as a preliminary step to computing $\zeta_{c,d,p}(s)$. See [39, Section 16.4] for a detailed exposition of this approach, and its connection with Higman's conjecture.

It follows easily from the structure theorem for finite abelian p-groups that

$$\zeta_{1,d,p}(s) = \zeta_p(s)\zeta_p(2s)\cdots\zeta_p(ds). \tag{3.8}$$

In [52] it is proved that, for all primes p, one has

$$\zeta_{2,2,p}(s) = \zeta_p(s)\zeta_p(2s)\zeta_p(3s)^2\zeta_p(4s).$$

No other explicit formulae of this kind are known.

4 Open problems and conjectures

In this section we collect a number of open problems and conjectures which we consider of central importance in the area of zeta functions of groups and rings.

4.1 Subring and subgroup zeta functions

Conjecture 4.1 ([22], p. 188). *Let $F_{c,d}$ denote the free class-c-nilpotent group on d generators. Then $\zeta_{F_{c,d}}(s)$ and $\zeta^\triangleleft_{F_{c,d}}(s)$ are almost uniform, i.e. there are rational functions $W_{c,d}(X,Y), W^\triangleleft_{c,d}(X,Y) \in \mathbb{Q}(X,Y)$ such that, for almost all primes p, one has*

$$\zeta_{F_{c,d},p}(s) = W_{c,d}(p, p^{-s}),$$
$$\zeta^\triangleleft_{F_{c,d},p}(s) = W^\triangleleft_{c,d}(p, p^{-s}).$$

Conjecture 4.2. *Let L be a class-c-nilpotent Lie ring of torsion-free rank n with upper central series $(Z_i(L))_{i \in \{0,\dots,c\}}$. Set $n_i := \mathrm{rk}(L/Z_i(L))$. Then, for almost all primes p, one has*

$$\deg_{p^{-s}}\left(\zeta^\triangleleft_{L,p}(s)\right) = -\sum_{i=0}^{c} n_i, \tag{4.1}$$

$$\lim_{s \to -\infty} (p^{-s})^{\sum_{i=1}^{c} n_i} \zeta^\triangleleft_{L,p}(s) = (-1)^n p^{\binom{n}{2}}. \tag{4.2}$$

Note that, for the primes p for which $\zeta_{L,p}(s)$ satisfies a functional equation of the form

$$\zeta_{L,p}^{\triangleleft}(s)|_{p \to p^{-1}} = (-1)^n p^{\binom{n}{2}-s} \sum_{i=0}^{c} n_i \zeta_{L,p}^{\triangleleft}(s), \qquad (4.3)$$

the equations (4.1) and (4.2) are simple corollaries of (4.3). In particular, Conjecture 4.2 holds if $c \le 2$; cf. [47, Theorem C]. For higher classes, however, it is known that equation (4.3) does not hold in general. All known examples nevertheless satisfy equations (4.1) and (4.2); see e.g. [21].

Problem 4.3. *Characterise nilpotent Lie rings for which the functional equation* (4.3) *holds for almost all primes p.*

A 'conjectural' characterisation has been given in [21, Chapter 4].

Problem 4.4. *Let G be the group defined in Example 3.2. Compute the local subgroup zeta functions $\zeta_{G,p}(s)$.*

Conjecture 4.5. *Let L be a class-2-nilpotent Lie ring with*

$$\mathrm{rk}(L/L') = d, \mathrm{rk}(Z(L)) = m, \mathrm{rk}(L/Z(L)) = r.$$

Let $\alpha^{\triangleleft}$ denote the abscissa of convergence of $\zeta_L^{\triangleleft}(s)$. Then

$$\alpha^{\triangleleft} = \max_{k \in \{1,\dots,m\}} \left\{ d, \frac{k(m+d-k)+1}{r+k} \right\}.$$

That $\alpha^{\triangleleft}$ is greater or equal to the right-hand side was proved in [42]. Equality has been proved for the free class-2-nilpotent groups $F_{2,d}$ in [50]. More generally, we pose the following problems:

Problem 4.6. *Given a ring L, determine the abscissae of convergence of its subring and ideal zeta functions, respectively.*

Problem 4.7. *Given a ring L, determine (a small superset of) the natural numbers a_i, b_i occurring in the denominators of its local (ideal) zeta functions; cf. Theorem 2.2.*

It follows from [17] that the abscissa of convergence of a ring's global zeta function is a simple function of these integers. Problem 4.7 is thus strictly harder than Problem 4.6. Even partial answers for specific families of rings as nilpotent or soluble Lie rings or 'simple' Lie rings like $\mathfrak{sl}_n(\mathbb{Z})$ would be very interesting.

4.2 Representation zeta functions

Problem 4.8 ([34], Problem 4.2). *Characterise rigid groups, and groups of polynomial representation growth (PRG).*

Problem 4.9. *Let G be a $\mathcal{T}$-group with representation zeta function $\zeta_G^{\mathrm{irr}}(s)$. Is the abscissa of convergence of $\zeta_G^{\mathrm{irr}}(s)$ a rational number? Does $\zeta_G^{\mathrm{irr}}(s)$ admit analytic continuation beyond its abscissa of convergence? Interpret the abscissa of convergence and the poles of the Euler factors of $\zeta_G^{\mathrm{irr}}(s)$ in terms of the structure of G.*

A positive answer to this problem would imply asymptotic statements about the numbers of twist-isoclasses of representations of $\mathcal{T}$-groups, analogous to Part (2) of Theorem 2.25.

Problem 4.10. *Let* $\mathrm{SL}_n^k(\mathbb{Z}_p) := \ker(\mathrm{SL}_n(\mathbb{Z}_p) \to \mathrm{SL}_n(\mathbb{Z}/(p^k\mathbb{Z})))$ *denote the kth congruence subgroup of* $\mathrm{SL}_n(\mathbb{Z}_p)$. *How do the functions* $\zeta^{\mathrm{irr}}_{\mathrm{SL}_n^k(\mathbb{Z}_p)}(s)$ *vary with the prime p? What are the abscissae of convergence of the zeta functions* $\zeta^{\mathrm{irr}}_{\mathrm{SL}_n^k(\mathbb{Z})}(s)$? *What about other 'classical' p-adic analytic groups?*

Problem 4.11. *Let G be an arithmetic group satisfying the CSP. Does its representation zeta function* $\zeta^{\mathrm{irr}}_G(s)$ *admit analytic continuation beyond its rational abscissa of convergence?*

Again, a positive answer would give us control over the asymptotic of the numbers $r_m(G)$ as m tends to infinity.

5 Exercises

In this section we collect a number of exercises. Harder exercises are marked with $\star$.

Exercise 5.1. Let q be a prime power, $n \in \mathbb{N}$, and $I \subseteq \{1, \ldots, n-1\}$. Show that the number of flags of type I in $\mathbb{F}_p^n$ is equal to $\binom{n}{I}_q$.

Exercise 5.2. Prove equation (2.12) directly.

Exercise 5.3. (cf. Example 3.1) Let p be a prime, and $L_p = L \otimes_{\mathbb{Z}} \mathbb{Z}_p$, where L is the Heisenberg Lie ring. Let $\Gamma = \mathrm{GL}_3(\mathbb{Z}_p)$ and $M \in \mathrm{Mat}_3(\mathbb{Z}_p)$. In the setup of Section 2.5, show that a coset ΓM corresponds to an ideal if and only if $M_{33} \mid M_{11}$ and $M_{33} \mid M_{22}$. Deduce that, for all primes

$$\zeta^{\triangleleft}_{L_p}(s) = \sum_{H \triangleleft_f L_p} |L_p : H|^{-s} = \zeta_p(s)\zeta_p(s-1)\zeta_p(3s-2).$$

Exercise 5.4 ($\star$). Let L be a ring of additive torsion-free rank n. Using the setup and notation of Section 2.3, show that a matrix $M = (M_{ij}) \in \mathrm{Tr}_3(\mathbb{Z}_p)$ encodes the generators of an ideal if and only if

$$\forall i \in \{1, \ldots, n\} : \ D\alpha^{-1}\mathfrak{R}(\alpha[i]) \equiv 0 \mod D_{ii},$$

(This is the 'ideal'-analogue of equation (2.23).)

Exercise 5.5 ($\star$). For $n \in \mathbb{N}$, let $L(n) = \mathbb{Z}^n$, considered as a ring with component-wise multiplication. Show that

$$\zeta_{L(2)}(s) = \zeta(s)^3\zeta(3s-1)\zeta(2s)^{-2}$$

Can you compute $\zeta_{L(n)}(s)$ for $n \geq 3$? Show that, for all $n \in \mathbb{N}$ and all primes p

$$\zeta_{L(n),p}(s) = \zeta_p(s)^n.$$

(A formula for $\zeta_{L(3),p}(s)$ can be found in [35, Proposition 6.3]. No formula seems to be known for $n \geq 4$.)

Exercise 5.6 ($\star$). Let G be the group defined in Example 3.2. Show that, for $p \neq 2$

$$\zeta_{G,p}^{\mathrm{irr}}(s) = W_1(p, p^{-s}) + b(p)W_2(p, p^{-s}),$$

where

$$W_1(X_1, X_2) = \frac{1 - X_2^3}{1 - X_1^3 X_2^3}, \quad W_2(X_1, X_2) = \frac{(X_1 - 1)(X_2 - 1)X_2^2}{(1 - X_1^2 X_2^2)(1 - X_1^3 X_2^3)}$$

and $b(p)$ is defined as in Example 2.15. Deduce the assertion of Theorem 3.5 in these cases.

Exercise 5.7. Establish formula (3.8).

References for Chapter III

[1] T. M. Apostol, *Introduction to analytic number theory. Undergraduate Texts in Mathematics.* Springer-Verlag, New York, 1976.

[2] N. Avni, Arithmetic groups have rational representation growth, arXiv:0803.1331, 2008.

[3] N. Avni, B. Klopsch, U. Onn and C. Voll, On representation zeta functions of groups and a conjecture of Larsen–Lubotzky. *C. R. Math. Acad. Sci. Paris*, 348, (7–8), 363–367, 2010.

[4] N. Avni, B. Klopsch, U. Onn and C. Voll, Representation zeta functions for SL_3. In preparation, 2010.

[5] N. Avni, B. Klopsch, U. Onn and C. Voll, Representation zeta functions of compact p-adic analytic groups and arithmetic groups. arXiv:1007.2900, 2010.

[6] M. N. Berman, Uniformity and functional equations for zeta functions of $\mathfrak{K}$-split algebraic groups. arXiv:0802.4207, to appear in *Amer. J. Math.*, 2010.

[7] A. I. Borevich and I. R. Shafarevich, *Number theory*, translated from the Russian by Newcombe Greenleaf, *Pure and Applied Mathematics*, vol. 20, Academic Press, New York and London, 1966.

[8] C. W. Curtis and I. Reiner, *Methods of representation theory, with applications to finite groups and orders*, vol. 1, John Wiley & Sons, 1981.

[9] J. Denef, Report on Igusa's local zeta function, *Séminaire Bourbaki*, 43 (201–203), 359–386, 1990–91.

[10] M. P. F. du Sautoy, Zeta functions of groups and rings: uniformity, *Israel J. Math.*, 86, 1–23, 1994.

[11] M. P. F. du Sautoy, Finitely generated groups, *p*-adic analytic groups and Poincaré series, *Ann. of Math.* (2), 137 (3), 639–670, 1993.

[12] M. P. F. du Sautoy, Counting p-groups and nilpotent groups, *Publ. Math. I.H.E.S.* 92, 63–112, 2000.

[13] M. P. F. du Sautoy, A nilpotent group and its elliptic curve: non-uniformity of local zeta functions of groups, *Israel J. Math.*, 126, 269–288, 2001.

[14] M. P. F. du Sautoy, Counting subgroups in nilpotent groups and points on elliptic curves, *J. Reine Angew. Math.*, 549, 1–21, 2002.

[15] M. P. F. du Sautoy, *Zeta functions of groups: The quest for order versus the flight from ennui*, Groups St. Andrews 2001 in Oxford, London Math. Soc. Lecture Note Ser., 304, Cambridge University Press, 2003, pp. 150–189.

[16] M. P. F. du Sautoy, Counting conjugacy classes, *Bull. London Math. Soc.* 37 (1), 37–44, 2005.

[17] M. P. F. du Sautoy and F. J. Grunewald, Analytic properties of zeta functions and subgroup growth, *Ann. of Math.* 152, 793–833, 2000.

[18] M. P. F. du Sautoy and A. Lubotzky, Functional equations and uniformity for local zeta functions of nilpotent groups, *Amer. J. Math.*, 118 (1), 39–90, 1996.

[19] M. P. F. du Sautoy and D. Segal, *Zeta functions of groups*, New horizons in pro-*p* groups, Progr. Math., Birkhäuser Verlag, Boston MA, 2000, pp. 249–286.

[20] M. P. F. du Sautoy and G. Taylor, The zeta function of $\mathfrak{sl}_2$ and resolution of singularities, *Math. Proc. Cambridge Philos. Soc.*, 132 (1), 57–73, 2002.

[21] M. P. F. du Sautoy and L. Woodward, *Zeta functions of groups and rings*, Lecture Notes in Mathematics, vol. 1925, Springer-Verlag, Berlin, 2008.

[22] F. J. Grunewald, D. Segal, and G. C. Smith, Subgroups of finite index in nilpotent groups, *Invent. Math.*, 93, 185–223, 1988.

[23] G. H. Hardy and M. Riesz, *The general theory of Dirichlet's series*, Cambridge Tracts in Mathematics and Mathematical Physics, No. 18, Cambridge University Press, 1915.

[24] H. Hironaka, Resolution of singularities of an algebraic variety over a field of characteristic zero. *I, II, Ann. of Math.* (2) 79, 109–203, 1964; ibid. (2) 79, 205–326, 1964.

[25] R. E. Howe, On representations of discrete, finitely generated, torsion-free, nilpotent groups, *Pacific J. Math.*, 73 (2), 281–305, 1977.

[26] E. Hrushovski and B. Martin, *Zeta functions from definable equivalence relations*, math.LO/0701011 on arxiv.org, 2007.

[27] J. E. Humphreys, *Reflection groups and Coxeter groups*, Cambridge Studies in Advanced Mathematics, vol. 29, Cambridge University Press, Cambridge, 1990.

[28] J.-I. Igusa, *An introduction to the theory of local zeta functions*, AMS/IP Studies in Advanced Mathematics, vol. 14, American Mathematical Society, Providence, RI, 2000.

[29] K. Ireland and M. Rosen, *A classical introduction to modern number theory*, GTM 84, Springer-Verlag, 1982.

[30] A. Jaikin-Zapirain, Zeta function of representations of compact p-adic analytic groups, *J. Amer. Math. Soc.*, 19, (19), 91–118, 2006.

[31] B. Klopsch, Zeta functions related to the pro-p group $SL_1(\Delta_p)$, *Math. Proc. Cambridge Philos. Soc.*, 135, 45–57, 2003.

[32] B. Klopsch and C. Voll, Zeta functions of three-dimensional p-adic Lie algebras, *Math. Z.*, 263 (1), 195–210, 2009.

[33] B. Klopsch and C. Voll, Igusa-type functions associated to finite formed spaces and their functional equations, *Trans. Amer. Math. Soc.*, 361 (8), 4405–4436, 2009.

[34] M. Larsen and A. Lubotzky, Representation growth of linear groups, *J. Eur. Math. Soc.* (JEMS), 10 (2), 351–390, 2008.

[35] R. I. Liu, Counting subrings of $\mathbb{Z}^n$ of index k, *J. Combin. Theory Ser. A*, 114, 278–299, 2007.

[36] A. Lubotzky and A. R. Magid, Varieties of representations of finitely generated groups, *Mem. Amer. Math. Soc.*, 58 (336), xi+117 pp, 1985.

[37] A. Lubotzky, A. Mann, and D. Segal, Finitely generated groups of polynomial subgroup growth, *Israel J. Math.*, 82 (1-3), 363–371, 1993.

[38] A. Lubotzky and B. Martin, Polynomial representation growth and the congruence subgroup growth, *Israel J. Math.*, 144, 293–316, 2004.

[39] A. Lubotzky and D. Segal, *Subgroup growth*, Birkhäuser Verlag, 2003.

[40] L. Manivel, *Symmetric Functions, Schubert Polynomials and Degeneracy Loci*, SMF/AMS Texts and Monographs, Volume 6, SMF/AMS, 2001.

[41] C. Nunley and A. R. Magid, Simple representations of the integral Heisenberg group, *Contemp. Math.*, 82, 89–96, 1898.

[42] P. M. Paajanen, On the degree of polynomial subgroup growth in class-2-nilpotent groups, *Israel J. Math.*, 157, 323–332, 2007.

[43] D. Segal, *Subgroups of finite index in soluble groups. I*, Proceedings of groups – St. Andrews 1985 (Cambridge), London Math. Soc. Lecture Note Ser., vol. 121, Cambridge Univ. Press, 1986, pp. 307–314.

[44] T. A. Springer, *Linear algebraic groups*, second ed., Progress in Mathematics, vol. 9, Birkhäuser Boston Inc., Boston, MA, 1998.

[45] R. P. Stanley, *Combinatorics and commutative algebra*, Birkhäuser, 1996, second edition.

[46] R. P. Stanley, *Enumerative combinatorics, Cambridge Studies in Advanced Mathematics*, 49, vol. 1, Cambridge University Press, 1997.

[47] C. Voll, *Functional equations for zeta functions of groups and rings*, Ann. of Math., 172, 1181–1218, 2010.

[48] C. Voll, Zeta functions of groups and enumeration in Bruhat-Tits buildings, *Amer. J. Math.*, 126, 1005–1032, 2004.

[49] C. Voll, Functional equations for local normal zeta functions of nilpotent groups, *Geom. Func. Anal.* (GAFA), 15, 274–295, 2005, with an appendix by A. Beauville.

[50] C. Voll, Normal subgroup growth in free class-2-nilpotent groups, *Math. Ann.*, 332, 67–79, 2005.

[51] C. Voll, Counting subgroups in a family of nilpotent semidirect products, *Bull. London Math. Soc.*, 38, 743–752, 2006.

[52] C. Voll, Enumerating finite class-2-nilpotent groups on 2 generators, *C. R. Math. Acad. Sci. Paris*, 347 (23-24), 1347–1350, 2009.

[53] J. Wlodarczyk, Simple Hironaka resolution in characteristic zero, *J. Amer. Math. Soc.*, 18 (4), 779–822, 2005.

Index

Printed in the United States
by Baker & Taylor Publisher Services